Neuro-Behçet's Disease

Erdem Tüzün • Murat Kürtüncü
Editors

Neuro-Behçet's Disease

Pathogenesis, Clinical Aspects, Treatment

Springer

Editors
Erdem Tüzün
Istanbul University
Istanbul, Turkey

Murat Kürtüncü
Istanbul University
Istanbul, Turkey

ISBN 978-3-030-55272-5 ISBN 978-3-030-55273-2 (eBook)
https://doi.org/10.1007/978-3-030-55273-2

This Springer imprint is published by the registered company Springer Nature Switzerland AG
The registered company address is: Gewerbestrasse 11, 6330 Cham, Switzerland

Preface

This book is about a puzzling disease that was first described in 1937 by a Turkish dermatologist, Hulusi Behçet. He used the term triple symptom complex consisting of aphthous ulcers in the mouth and genitalia, and uveitis. Behçet's disease may involve many organs including the nervous system. Neurological involvement is termed as neuro-Behçet's disease that occurs in 5–10% of patients.

Although many years have passed after its first description, the pathogenesis of Behçet's disease still conserves its enigma. Developments in immunology and clinical trials on Behçet's disease in the last 15 years facilitated our understanding of the pathogenesis of Behçet's disease. However, the management of neuro-Behçet's disease still relies upon data derived from case studies or studies performed on patients with systemic involvement.

This book aims to fill a gap caused by the absence of a well-structured textbook on neuro-Behçet's disease. With this book, we hope to ease the clinical decision making for clinicians who deal with neuro-Behçet's disease patients.

First editions of any scientific text may contain errors and this book is no exception. We welcome suggestions and criticism from readers.

We would like to dedicate this book to patients with this disabling chronic disease with the hope that their management may be facilitated by this book.

We wish to express our gratitude to the authors; each have immense expertise on this filed. We also owe much to Hande Yüceer and the editorial staff at Springer: they undertook a demanding task of collecting chapters.

We are indebted to our families and colleagues who encouraged and supported us during the implementation of this book.

Istanbul, Turkey

Erdem Tüzün
Murat Kürtüncü

Contents

Neuro-Behçet's Disease: Epidemiology and Genetics

Hande Yüceer and Erdem Tüzün

Introduction

Behçet's disease (BD) is an inflammatory disorder with multisystemic clinic manifestations including recurrent oral aphthae, erythema nodosum, genital ulcerations, eye lesions (uveitis, retinal vasculitis), and positive pathergy test [1, 2].

In addition to these symptoms, nervous system has been reported to be involved in nearly 5–10% of the BD patients, which refers to the neuro-Behçet's disease (NBD) [3]. Neurological involvement mostly occurs after a few years from the onset of BD but with more severe outcomes. Therefore, NBD has been reported as the main mortality cause of BD [4]. It is mainly seen in two main forms: parenchymal involvement of the central nervous system (CNS) and thrombosis of the cerebral venous sinuses [1, 5]. The typical case of CNS parenchymal involvement is generally due to brainstem and diencephalon lesions and is characterized by behavioral change, dysarthria, hemiparesis, and ataxia [6]. Also, in cerebral venous sinus thrombosis (CVST), increased intracranial pressure, seizures, and headache can be encountered [1, 6]. Even though NBD is infrequent in many regions of the world, it is highly cross-referred in the context of distinguishing from other demyelinating and inflammatory disorders of the CNS [5].

While the highest prevalence of BD is on the ancient Silk Road, which links the Far East and the Mediterranean region, it is uncommon in North America and Europe [5]. Hence, BD has the highest incidence in countries like Turkey, Korea, China, and Japan [7, 8].

The etiologies of both BD and NBD are still unclear, but immune system and genetic defects seem to play a causal role in the etiopathogenetic mechanisms of the disorder [6, 9]. In this section, the current epidemiology and genetic features such as familial aggregation, MHC class genes, genome-wide association studies

H. Yüceer (✉) · E. Tüzün
Department of Neuroscience, Aziz Sancar Institute of Experimental Medicine, Istanbul University, Istanbul, Turkey

E. Tüzün, M. Kürtüncü (eds.), *Neuro-Behçet's Disease*,
https://doi.org/10.1007/978-3-030-55273-2_1

(GWAS), polymorphisms, and epigenetic studies with regard to both BD and NBD will be reviewed.

Epidemiology

Prevalence of Neuro-Behçet's Disease

BD has a global dispersion. Despite that, it is detected more often in the ancient trade route (Silk Road), along the path of which, Middle East, Far East, and Mediterranean countries are located [8, 10]. In the Silk Road countries, every 14–20 people among 100,000 is expected to suffer from BD [10]. The prevalence of BD in Turkey ranges from 20 to 602/100000, which is considered the most frequent in the world [8]. Even only in Istanbul, as a representation of Turkey profile alone, the frequency rate reaches 421 per 100,000 people [9]. Israel (50–185/100000), Iran (16.7–100/100000), Iraq (17/100000), Japan (7–13.5/100000), and China (2.62/100000) are the other countries where BD is relatively more common [9–11]. In terms of neurological symptoms, Caucasians (23%) have been observed to suffer more from NBD compared to Middle Eastern populations (3–10%) [12].

On the other hand, the prevalence of NBD in BD is only between 3% and 9% [13]. However, a postmortem study from Japan has shown that 20% of BD cases had neurological pathologies, suggesting that NBD prevalence is higher than clinically anticipated [14]. Additionally, the parenchymal involvement of CNS in BD has been claimed to be more prevalent compared to other forms of NBD [13]. In a long-term follow-up study of Kural-Seyahi et al. [15], the mortality rate has been found to be 9.8% for 42 out of 387 BD patients with neurological manifestations.

Through the last decades, the studies related to the epidemiology of NBD have expanded due to the increased rate of NBD diagnosis in neurological practice. The studies in the literature about BD conflict with each other regarding the percentage of the neurological manifestations of the disease. The frequency of NBD in hospital-based research studies dramatically varies in the range between 1.3% and 59%, which may be due to diverse study designs, the inconsistency in definition of clinical features of NBD, and the variance of geoepidemiology of the disease [15].

Age

The mean age of onset of BD was around 25, while the mean age of onset of NBD was found to be at around 31 in a study from a single Behçet's disease outpatient clinic [13]. Likewise, in other studies, the mean age of onset of NBD has been shown to range between 25 and 34 [5] (Table 1). Nevertheless, BD and NBD have been reported in children and juveniles, as well [16–18]. However, it is hard to

Table 1 Demographic features of BD and NBD patients [5]

Location			Age of onset	
	Male	Female	BD	NBD
Tunisia	61	14	31	34
Saudi Arabia	42	10	27	–
Turkey	155	45	25·8	31·5
Korea	9	12	–	35·1
West Indies	5	2	–	35·5
France	78	31	–	32
Iraq	14	6	–	34·1
Iran	15	3	–	34·7
Italy	7	20	27·7	34·7
UK	11	11	–	30

investigate the frequency of the disease in pediatric population due to the uncertainty of neurological symptoms and the unavailability of broad sample size. The differential diagnosis of NBD and the determination of negative/positive presumptions regarding other neurological disorders are critical especially after 40 years of age [19].

Gender

Although the gender differences have not been reported to be significant in BD, the overall outcome of the studies shows that men tend to have NBD almost three times more than women [5, 13, 15]. However, severity of neurological symptoms is not associated with gender [13].

Genetic Factors

Familial Aggregation

BD's higher prevalence in specific regions indicates the involvement of genetic and environmental factors in the etiopathogenesis of the disease. Even though it is not an epidemic disorder, the occurrence of familial BD is prominent. The higher rates of hereditary distribution of BD are disclosed in various races as Turkish (18.2%), Korean (15.4%), and Jewish (13.2%). On the other hand, Chinese (2.6%), Japanese (2.2%), and various European populations (0–4.5%) have been reported to procure lower familial aggregation. Also, it has been detected that hereditability of BD increases when the disease gives its first symptoms in early ages [9].

Major Histocompatibility Complex (MHC).

In autoimmune disorders, the major histocompatibility complex (MHC) or human leukocyte antigen (HLA) complex is the most vigorous heterogenous genetic zone associated with initiation of T-cell response and antigen recognition. The high polymorphism rates of MHC genes do not only contribute to the diversification of populations but also may predispose individuals to autoimmune and autoinflammatory diseases like BD [20].

HLA-B51

The most directly colligated element of risk explaining the relationship between Silk Road and BD is HLA-B51 placed in chromosome 6p. Some of the early studies on the significance of HLA-B51 have arisen from Japan, where the prevalence of HLA-B51 has been reported as 57% [21]. In Mediterranean and Middle East regions, HLA-B51 is even more common and, for example, is found in 84% of BD cases in Turkey [22].

Non-carriers of HLA-B51 are less likely to develop BD in comparison to HLA-B51-positive cases [8]. HLA-B51's rate of predictability for BD has been calculated as 59% in a study conducted with 1215 BD patients and 1278 healthy controls [9]. While HLA-B51 is also associated with NBD, its prognostic significance is somewhat controversial. HLA-B51 positivity has been reported as high as 90% in an early study from Japan.

[23]. HLA-B51 might also serve in differentiation of NBD from other inflammatory disorders such as neuro-Sweet disease, which is strongly associated with HLA-Cw1 and B54 [24]. Although HLA-B51-allele-carrying NBD patients were reported to have worse prognosis by some authors, others suggested that HLA-B51 did not have a prognostic role in NBD [25, 26]. The HLA-B5103 allele has been indicated as a risk factor for neurological manifestations in BD [27].

The 3D model of HLA-B5101 has been designed by Maenaka et al. [28]. The figure has been open for public domain and is obtained from http://commons.wikimedia.org/ (Fig. 1).

Other MHC Class I Genes

HLA-B51 is not the only MHC class I allele related with NBD, and there are other types of HLA and MHC class I genes, the importance of which has been described in NBD etiopathogenesis. In the current literature, HLA-B51, HLA-B15, HLA-B27 have been shown as the major regions indicating the risk for BD development, while the protective role of HLA-A03 and HLA-B49 regions against the emergence of BD has been found [29].

MHC class I polypeptide-related sequence A (MICA) is another genetic factor with a key role in generating a tendency toward BD development. MICA is a

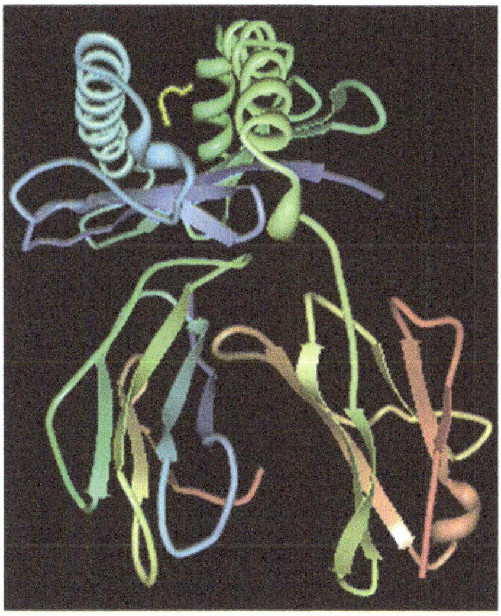

Fig. 1 BD- and NBD-associated MHC class I gene, HLA-B5101

functional gene positioned between the tumor necrosis factor and HLA-B genes [30]. The expression of this gene is increased during viral inflammation and stress. The expression of MICA is then followed by a cascade of autoimmune response cells encompassing NK cells, CD8+ T cells, and gamma delta T cells [9, 30, 31].

Non-MHC Complex Genes

With the contribution of genome-wide association studies (GWAS) and next-generation sequencing methods, non-MHC regions have been acknowledged as the probable cause of BD pointing out the multigenic nature of this disease [8].

Genome-Wide Association Studies

Genome-wide association studies (GWAS) have an efficient provision of impartial gene index to anticipate how different individuals gain susceptibility to complex diseases like BD through analysis of single-nucleotide polymorphisms (SNPs). Approximately 0.1% genetic variants in the human genome are SNPs [32]. Numerous common SNPs have been identified in various ethnic groups including Turkish, Iranian, Chinese, Japanese, and Korean BD patients [33] (Table 2).

Table 2 The genome-wide significant SNPs for the susceptibility of BD [33]

	SNP	Gene	Allele	Ethnicity
Mizuki et al. [34]	rs1495965	IL23R, IL12RB2	G	Japanese
Remmers et al. [35]	rs924080	IL23R, IL12RB2	A	Turkish, Japanese
Wu et al. [36] Remmers et al. [35]	rs1518111	IL10	A	Turkish
Mizuki et al. [34] Wu et al. [36]	rs1800871	IL10	T	Japanese
Li, et al. [37]	rs9494885	TNFAIP3	C	Han Chinese
Kirino et al. [38]	rs7574070	STAT4	A	Turkish
Hou, Yang, et al. [39]	rs897200	STAT4	A	Han Chinese
Kirino, et al. [38]	rs7616215	CCR1	T	Turkish
Qin et al. [40, 41]	rs13092160	CCR1, CCR3	T	Han Chinese
Kirino, et al. [38]	rs2617170	KLRC4	C	Turkish
Kirino et al. [38]	M694V	MEFV	V	Turkish
Kirino, et al. [38]	rs17482078	ERAP1	TT2	Turkish
Xavier et al. [42]	rs681343	FUT2	T	Iranian, Turkish
Kirino, et al. [38]	rs17810546	IL12A	A/G3	Turkish, mixed populations
Kirino et al. [38]	R381Q, G149R1	IL23R		Turkish, Japanese
Kirino et al. [38]	D299G, T399I1	TLR4		Turkish, Japanese
Kirino et al. [38]	R702W, G908RL1007fs1	NOD2		Turkish, Japanese

Il-10

Interleukin (IL)-10 is an anti-inflammatory cytokine intercepting the proinflammatory process by preventing the procurement of proinflammatory cytokines tumor necrosis factor (TNF)-α, IL-12, IL-6, IL-1, and interferon gamma (IFN-γ) [36]. Moreover, IL-10 is one of the first identified non-MHC complex genes implicated for BD susceptibility. Early GWAS from Turkey and Japan have shown the genome-wide significance of IL-10. Different variants of the IL-10 gene have been associated with BD. While the rs1518111 variant identified in Turkish BD patients is located in the intronic region, the rs1800871 and rs1800872 variants identified in Japanese patients are found in the promoter region of the IL-10 gene [34–36]. The majority of the SNP findings mentioned in Table 2 has been reproduced by other researchers. The rs1800872 variant has been replicated in Turkish and Korean patients; the rs1518111 variant has been reproduced in Iranian, Greek, British, and Korean patients; and the rs1800871 variant has been reproduced in Chinese patients [34–36]. It has been speculated that the above-mentioned SNPs lead to enhanced BD susceptibility through reduced IL-10 expression and subsequently increased tendency to proinflammatory responses [33]. The polymorphism of the IL-10 gene and its role in disease susceptibility have also been shown in NBD patients [43].

IL23R-IL12RB2

IL12R and IL23R gene regions have been widely studied in BD patients. Notably, IL-12 and IL-23 levels are enhanced in BD patients, and these two cytokines contribute to neutrophil, NK cell, and T-cell functions in inflammatory diseases [44–46]. The significance of the IL23R-IL12RB2 intergenic region in BD has been shown by several GWAS. In a GWA study from China including 421 controls and 407 BD patients, the interrelation between rs11209032, rs924080 SNPs and the IL23R-IL12RB2 locus was analyzed, and two IL23R-IL12RB region SNPs (rs11209032, rs924080) were detected in BD patients [40, 41]. In rs924080 risk allele-carrying individuals, the expression of IL23R is increased in peripheral blood cells, and LPS-induced proinflammatory cytokine production is enhanced [47]. Moreover, the GG genotype of the SNP rs17375018 in the IL23R gene was associated with BD, and in rs17375018 GG genotype carriers, increased serum IL-23 levels were linked to occurrence of neurological symptoms in BD [48]. Intriguingly, two missense SNPs of the IL23R gene (Gly149Arg in Japanese and Arg381Gln in Turkish cohorts) have been found to have lower prevalence in BD patient groups, suggesting that these variants have a protective action against BD development. The Arg381Gln variant appears to confer protection via reducing IL-23-dependent IL-17 production [38]. In many of these GWAS, although patients with NBD have been included, features of BD patients with and without neurological manifestations have not been signified.

IL12A

The IL12A gene encodes the IL-12 cytokine, which manages adaptive and innate immune responses [49]. The rs1780546 variant, which is located in the intergenic part of IL12A, has been suggested to have a connection with BD in a Turkish cohort [38]. That suggestion has been approved by a multi-ethnicity GWA study [50]. In addition, these results have been repeated in an Iranian GWAS with 973 BD patients (61 of them were NBD patients). It has been confirmed that IL12A-AS1 locus has a susceptibility role for BD [51]. The SNP rs1874886 was also defined as an assumed risk factor in another study for Spanish population [52].

ERAP1

Endoplasmic reticulum aminopeptidase-1 (ERAP1) is an enzyme having multiple functions for MHC class I alleles such as HLA-B51. The mutation of ERAP1 in HLA-B51-positive cases has pointed out the indicative role of ERAP1 in BD [53]. The fact that ERAP1 is a biomarker of BD only in HLA-B51-positive patients also supports the evident role of HLA-B51 in BD etiopathogenesis [33]. Furthermore, the link between ERAP1 and HLA-B51, and how interactions between these two molecules might cause BD through peptide binding differences, CD8+ cell activation, and NK cell inhibition, has been discussed in several studies [38, 54]. Although Kirino et al. also have mentioned the genome-wide significance of rs17482078, which is an ERAP1 variant, for BD risk in Turkey, the ample polymorphism for the same variant has not been found in Japanese patients [38].

STAT4

As an immune-related transcription factor, STAT4 (signal transducer and activator of transcription 4) is concerned with the formation of Th1/Th17 from T cells and the proinflammatory response initiated by IL-12, IL-17, and IL-23 [55]. The STAT4 variants are associated with alterations in gene expression patterns [56]. BD-related STAT4 variants are rs7574070 and rs897200. The rs7574070 variant has an impact on STAT4 gene expression. On the other hand, the rs897200 variant is involved not only in increased transcription of IL-17 but also augmented expression of the STAT4 gene. Interestingly, the rs897200 has also been reported to be detected in more severe BD cases [38, 39, 57].

TNFAIP3

The main function of tumor necrosis factor alpha-induced protein 3 (TNFAIP3) gene is to code ubiquitin-modifying enzyme A20 and to maintain innate immunity primarily by regularizing NF-kappa B and TNF activation [58, 59]. TNF-α is a major proinflammatory cytokine implicated in BD pathogenesis. TNF-inhibiting reagents have long been used for BD treatment in order to block Th17 differentiation and thus stop inflammation. TNFAIP3 gene rs9494885 polymorphism has demonstrated a genome-wide significance in the Chinese Han population, but this finding could not be replicated in European populations [29, 60, 61]. The impact of TNF-α polymorphism on disease predisposition has been demonstrated in NBD patients indicating the significance of this cytokine in NBD pathogenesis [43].

CCR1-CCR3

Chemotactic cytokine receptor 1 (CCR1) and chemotactic cytokine receptor 3 (CCR3) codify G-protein-coupled chemokine receptors and take part in inflammatory responses [62]. There has been relatively small number of research on the BD-indicative polymorphisms of CCR1 and CCR3 genes. A well-known GWAS has claimed that CCR1 and CCR3 loci confer susceptibility to BD [35]. These results have been replicated in a Han Chinese GWAS conducted with 1685 healthy controls and 653 BD patients. As a result, three SNPs of CCR1 and CCR3 genes (rs13084057, rs13092160, and rs13075270) have been found above genome-wide significance level [57]. However, none of 653 BD patients in the study showed neurological symptoms. In addition to these SNPs, rs7616215 has also been verified for the Turkish ethnicity [38]. These outcomes provide evidence that the SNPs (rs13084057, rs13092160, rs13075270, and rs7616215) located intergenically between CCR1 and CCR3 genes could have a pathogenic action in BD [38, 57].

KLR

The KLRK1, KLRC1, KLRC2, KLRC3, and KLRC4 genes, which encode killer cell lectin-like receptors, have been identified as susceptibility factors for BD in GWAS [38]. Especially, C allele of KLRC4 variant rs2617170 has been related with BD pathogenesis in Japanese and Turkish patients [63].

GIMAP

The family of GIMAP genes controls the encoding of GTP binding proteins especially in T cells [64]. The expression of GIMAP1 and GIMAP4 mRNAs was lower in the peripheral blood mononuclear cells (PBMC) of Japanese and Korean BD patients. 11.6% of these patients had central nervous system involvement. In the same GWAS, SNP rs1608157 was associated with increased BD risk [37]. Another GWA study from Iran, in which 6% of the BD patients had neurological manifestations, has also mentioned the role of SNP rs1916012 of GIMAP7 for BD susceptibility in Iran [51]. However, the results of these GWAS could not be replicated for other nationalities [50]. In this sense, GIMAP genes may be critical to evaluate the ethnic differentiation of T-cell mediation in BD pathogenesis.

FUT2

FUT2 gene manages the synthesis of ABO antigens through encoding of fucosyltransferase 2. These histo-blood group antigens are found in intestinal mucosa or bodily secretions, and lack of FUT2 activation may result in enhanced inflammatory responses [65]. The SNP rs681343 variant of FUT2 has been shown to have genome-wide significance for both BD and NBD patients [42]. The relationship between gut microbiome, autoimmunity, and BD etiology has yet to be investigated through FUT2.

CPLX-1

CPLX (complexin) genes basically mediate the release of neurotransmitters via stabilizing soluble N-ethylmaleimide-sensitive factor attachment protein receptor (SNARE) complex and thus have prominent impact on neuron function and survival [66]. The CPLX-1 rs936551 SNP has been specified in BD patients in a GWA study [35]. Further in another study, rs936551 SNP has been identified in both BD and NBD patients and has been associated with raised PBMC CPLX-1 expression levels [67]. The exact mechanism of action of CPLX-1 in inflammation is not known and needs to be determined by future studies.

Other Polymorphism Studies

The association between KIAA1529, CPVL, LOC100129342, and BD risk has been mentioned in a GWA study. In this study, 7.2% of the patients had neurological involvement. The genetic predisposition of the populations on the Silk Road has been explained by LOC100129342 polymorphism in the same article. The novel genetic susceptible ubiquitination pathway genes UBASH3B and UBAC2 loci have also been mentioned [68]. Similarly, SUMO4 (small ubiquitin-like modifier 4) has been evaluated as an indicator of BD risk in Chinese patients [69]. Moreover, the polymorphism of MBL2–54 gene A allele (mannose-binding lectin) in neurological symptoms and genital ulcer has been identified in another study [70]. Intercellular adhesion molecule 1 (ICAM-1) and vascular endothelial growth factor (VEGF) gene expressions have been linked to the inflammatory pathways of BD [71, 72].

Rare Non-synonymous Variants

BD-associated rare non-synonymous variants (NSVs) of the Toll-like receptor 4 gene (TLR4), the nucleotide-binding oligomerization domain 2 gene (NOD2), and the familial Mediterranean fever gene (MEFV) have also been analyzed by targeted next-generation sequencing methods [29]. TLR4, NOD2, and MEFV have been proposed to be related with innate immune response and inflammatory mechanisms in BD [38]. The low frequencies of D299G, T399I1 NSVs of TLR4, R702W, G908RL1007fs1 NSVs of NOD2, and the mutation of M694V variant of MEFV have been reported for BD susceptibility in Japan and Turkey [38, 53, 73]. Another meta-analysis has shown the mutation of M680I and M694V variants of MEFV as risk loci for Turkish BD patients; howcvcr E148Q NSV of MEFV was not found to be genomically significant [74]. In contrast, the heterozygous mutation of E148Q variant has been found in one NBD case with periodic fever symptoms [75].

Epigenetic Studies

The term "epigenetics" defines the modification in gene expression without DNA sequential change, and main epigenetic mechanisms are histone remodeling, DNA methylation, and miRNA-mediated gene expression regulation [43]. Epigenetic studies are believed to be a key to understand the link between distinct geographical localization and the genetic mechanisms of BD [48].

The role of activated interspersed repetitive sequences (IRSs) like Alu and LINE-1 resulting from hypomethylation has been recently investigated for autoimmune disorders including BD. Mainly, extracted DNA from PBMCs of BD patients has been analyzed, and the alteration in the IRS methylation has been found as a possible risk factor for BD [29, 76].

The relation of several miRNAs with NBD's inflammatory component has also been shown. Firstly, the expression of miR155 in dendrites was found significantly lower in BD patients compared to healthy controls. Whether this contributes to induction of neurological symptoms in BD needs to be determined. Secondly, the decreased expression of miR23b, miR196, miR155, miR638, and miR4488 in CD4 T cells and the increased expression of miR21, miR182, and miR3591 have been observed in BD patients. These up- and downregulations of miRNAs may have an impact on immune-related gene functions [48, 77]. Also, in another study, decreased PBMC mRNA expression level of miR185, the regulator of CPLX1 expression, has been shown in both BD and NBD patients [67].

Conclusion

The etiology of BD with neurological involvement is complex and yet to be discovered. However, the role of both innate and adaptive immune systems (depending on Th1/Th17 differentiation) as well as genetic susceptibilities (mainly HLA associated) has been shown in recent studies. Since the majority of the researches have investigated BD, our knowledge on NBD regarding epidemiologic and genetic data is limited. The lower prevalence of NBD compared to BD and the late onset of neurological findings in BD are factors making the investigation of NBD etiology harder to achieve [13]. Nevertheless, due to shared immunological and genetic mechanisms, the results of the BD studies might be expected to be correlated with NBD studies in the future. Even so, these findings should further be replicated with patients having neurological symptoms, and the results need to be verified for NBD.

References

1. Eraksoy M, Akman-Demir G. Merkezi sinir sisteminin miyelin hastalıkları. In: Nöroloji. 2nd ed. İstanbul: Nobel Tıp Kitabevleri; 2015. p. 626–7. Retrieved from http://www.itfnoroloji.org/MS/MS.htm.
2. International Study Group for Behçet's Disease. Criteria for diagnosis of Behcet's disease. Lancet. 1990;335(8697):1078–80. https://doi.org/10.1016/0140-6736(90)92643-V.
3. Akman-Demir G, Saip S, Siva A. Behçet's disease. Curr Treat Options Neurol. 2011;13(3):290–310. https://doi.org/10.1007/s11940-011-0120-2.
4. Saadoun D, Wechsler B, Desseaux K, Le Thi Huong D, Amoura Z, Resche-Rigon M, Cacoub P. Mortality in Behçet's disease. Arthritis Rheum. 2010;62(9):2806–12. https://doi.org/10.1002/art.27568.
5. Al-Araji A, Kidd DP. Neuro-Behçet's disease: epidemiology, clinical characteristics, and management. Lancet Neurol. 2009;8(2):192–204. https://doi.org/10.1016/S1474-4422(09)70015-8.
6. Haghighi AB, Pourmand R, Nikseresht AR. Neuro-Behçet disease: a review. Neurologist. 2005;11(2):80–9. https://doi.org/10.1097/01.nrl.0000156343.16797.c4.
7. Kaufman WS, Mcnamara EK, Jorizzo JL. Behçet's disease. In: Kelley's textbook of rheumatology, vol. 9(93); 2012. p. 1589. https://doi.org/10.1016/B978-1-4377-1738-9.00093-1.
8. Verity DH, Wallace GR, Vaughan RW, Stanford MR. Behçet's disease: from Hippocrates to the third millennium. Br J Ophthalmol. 2003;87(9):1175–83.
9. Ozlu E, Uzuncakmak T. Epidemiology and genetic aspects of Behçet's disease. In: Behcet's disease – a compilation of recent research and review studies: SM Group; 2017. p. 1–8.
10. Oguz I, Hizli P, Gonul M. The epidemiology of Behçet's disease. In: Behcet's disease: InTech; 2017. p. 15–26. https://doi.org/10.5772/68055.
11. Gönül M, Kartal SP. Introductory chapter: Behcet's disease – an overview. In: Behcet's disease: InTech; 2017. https://doi.org/10.5772/intechopen.69050.
12. Joseph FG, Scolding NJ. Neuro-Behçet's disease in Caucasians: a study of 22 patients. Eur J Neurol. 2007;14(2):174–80. https://doi.org/10.1111/j.1468-1331.2006.01613.x.
13. Saip S, Akman-Demir G, Siva A. Neuro-Behcet syndrome. Handb Clin Neurol. 2014;121:1703–23. https://doi.org/10.1016/b978-0-7020-4088-7.00110-3.
14. Lakhanpal S, Tani K, Lie JT, Katoh K, Ishigatsubo Y, Ohokubo T. Pathologic features of Behçet's syndrome: a review of Japanese autopsy registry data. Hum Pathol. 1985;16(8):790–5. https://doi.org/10.1016/S0046-8177(85)80250-1.
15. Kural-seyahi E, Fresko I, Seyahi N, Ozyazgan Y, Mat C, Hamuryudan V, Yurdakul S, Yazici H. The long-term mortality and morbidity of Behçet syndrome: a 2-decade outcome survey of 387 patients followed at a dedicated center. Medicine (Baltimore). 2003;82(1):60–76.
16. Chessa Ricotti G, De Martino M, Martini R, Rossi ME, Pietraperzia M, Vierucci A. A case of Behcet's disease in a child: clinical, genetic, and immunologic characteristics. Pediatr Med Chir. 1983;5(4):211–8. Retrieved from http://www.ncbi.nlm.nih.gov/pubmed/6647083.
17. Koné-Paut I. Behçet's disease in children, an overview. Pediatr Rheumatol. 2016;14(1):10. https://doi.org/10.1186/s12969-016-0070-z.
18. Mora P, Menozzi C, Orsoni JG, Rubino P, Ruffini L, Carta A. Neuro-Behçet's disease in childhood: a focus on the neuro-ophthalmological features. Orphanet J Rare Dis. 2013;8:18. https://doi.org/10.1186/1750-1172-8-18.

19. Sefidbakht S, Babaeinejad M, Jali R, Zare Z, Sahraian A, Safari A, Haghighi AB. The McDonald criteria for dissemination in space in the differential diagnosis of multiple sclerosis and neuro-Behcet's disease. Neurol Asia. 2014;19(2):47–52.
20. Yalçin B. Major histocompatibility complex (MHC) molecules: their common characteristics and relations with diseases. Turkderm Deri Hastaliklari ve Frengi Arsivi. 2013;47(1):12–7. https://doi.org/10.4274/turkderm.47.s2.
21. Kidd D, Steuer A, Denman AM, Rudge P. Neurological complications in Behçet's syndrome. Brain. 1999;122(11):2183–94.
22. Akman-Demir G, Serdaroglu P, Tasci B. Clinical patterns of neurological involvement in Behçet's disease : evaluation of 200 patients. Brain. 1999;122:2171–81.
23. Shinkeigaku R. Central nervous system involvement in Behçet's disease. Rinsho Shinkeigaku. 2001;41(12):1147–9.
24. Hisanaga K. Neuro-neutrophilic disease: neuro-Behçet disease and neuro-sweet disease. Intern Med. 2007;46(4):153–4. https://doi.org/10.2169/internalmedicine.46.0175.
25. Esatoglu SN, Kutlubay Z, Ucar D, Hatemi I, Uygunoglu U, Siva A, Hatemi G. Behçet's syndrome: providing integrated care. J Multidiscip Healthc. 2017;10:309–19. https://doi.org/10.2147/JMDH.S93681.
26. Noel N, Bernard R, Wechsler B, Resche-Rigon M, Depaz R, Le Thi Huong Boutin D, Piette JC, Drier A, Dormont D, Cacoub P, Saadoun D. Long-term outcome of neuro-behçet's disease. Arthr Rheumatol. 2014;66(5):1306–14. https://doi.org/10.1002/art.38351.
27. Demirseren D, Ceylan G, Akoglu G, Emre S, Erten S, Arman A, Metin A. HLA-B51 subtypes in Turkish patients with Behçet's disease and their correlation with clinical manifestations. Genet Mol Res. 2014;13(133):4788–96. https://doi.org/10.4238/2014.July.2.8.
28. Maenaka K, Maenaka T, Tomiyama H, Takiguchi M, Stuart DI, Jones EY. Nonstandard peptide binding revealed by crystal structures of Hla-B5101 complexed with HIV immunodominant epitopes. J Immunol. 2000;165:3260. https://doi.org/10.2210/PDB1E28/PDB.
29. Hatemi G, Seyahi E, Fresko I, Talarico R, Hamuryudan V. One year in review 2016: Behçet's syndrome. Clin Exp Rheumatol. 2016;102:10–22.
30. Wei F, Zhang Y, Li W. A meta-analysis of the association between Behçet's disease and MICA-A6. Biomed Rep. 2016;4(6):741–5. https://doi.org/10.3892/br.2016.644.
31. Lee YH, Song GG. Associations between major histocompatibility complex class I chain-related gene A polymorphisms and susceptibility to Behcet's disease. A meta-analysis. Z Rheumatol. 2015;74(8):714–21. https://doi.org/10.1007/s00393-014-1536-3.
32. Jorde LB, Wooding SP. Genetic variation, classification and "race". Nat Genet. 2004;36(11s):S28–33. https://doi.org/10.1038/ng1435.
33. Takeuchi M, Kastner DL, Remmers EF. The immunogenetics of Behçet's disease: a comprehensive review. J Autoimmun. 2015;64:137–48. https://doi.org/10.1016/j.jaut.2015.08.013.
34. Mizuki N, Meguro A, Ota M, Ohno S, Shiota T, Kawagoe T, Ito N, Kera J, Okada E, Yatsu K, Song YW, Lee EB, Kitaichi N, Namba K, Horie Y, Takeno M, Sugita S, Mochizuki M, Bahram S, Ishigatsubo Y, Inoko H. Genome-wide association studies identify IL23R-IL12RB2 and IL10 as Behçet's disease susceptibility loci. Nat Genet. 2010;42(8):703–6. https://doi.org/10.1038/ng.624.
35. Remmers EF, Cosan F, Kirino Y, Ombrello MJ, Abaci N, Satorius C, Le JM, Yang B, Korman BD, Cakiris A, Aglar O, Emrence Z, Azakli H, Ustek D, Tugal-Tutkun I, Akman-Demir G, Chen W, Amos CI, Dizon MB, Akdag Kose A, Azizlerli G, Erer B, Brand OJ, Kaklamani VG, Kaklamanis P, Ben-Chetrit E, Stanford M, Fortune F, Ghabra M, Ollier WER, Cho YH, Bang D, O'Shea J, Wallace GR, Gadina M, Kastner DL, Gül A. Genome-wide association study identifies variants in the MHC class I, IL10, and IL23R-IL12RB2 regions associated with Behçet's disease. Nat Genet. 2010;42(8):698–702. https://doi.org/10.1038/ng.625.
36. Wu Z, Zheng W, Xu J, Sun F, Chen H, Li P, Chen S, Shen M, Zhang W, You X, Wu Q, Zhang F, Li Y. IL10 polymorphisms associated with Behçet's disease in Chinese Han. Hum Immunol. 2014;75(3):271–6. https://doi.org/10.1016/j.humimm.2013.11.009.

37. Lee YJ, Horie Y, Wallace GR, Choi YS, Park JA, Choi JY, Park JA, Choi JY, Song R, Kang YM, Kang SW, Baek H, Kitaichi N, Meguro A, Mizuki N, Namba K, Ishida S, Kim J, Niemczyk E, Lee EY, Song YW, Ohno S, Lee EB. Genome-wide association study identifies GIMAP as a novel susceptibility locus for Behçet's disease. Ann Rheum Dis. 2013;72(9):1510–6. https://doi.org/10.1136/annrheumdis-2011-200288.
38. Kirino Y, Zhou Q, Ishigatsubo Y, Mizuki N, Tugal-Tutkun I, Seyahi E, Özyazgan Y, Ugurlu S, Erer B, Abaci N, Ustek D, Meguro A, Ueda A, Takeno M, Inoko H, Ombrello MJ, Satorius CL, Maskeri B, Mullikin JC, Sun HW, Gutierrez-Cruz G, Kim Y, Wilson AF, Kastner DL, Gül A, Remmers EF. Targeted resequencing implicates the familial Mediterranean fever gene MEFV and the toll-like receptor 4 gene TLR4 in Behcet disease. Proc Natl Acad Sci. 2013;110(20):8134–9. https://doi.org/10.1073/pnas.1306352110.
39. Hou S, Yang Z, Du L, Jiang Z, Shu Q, Chen Y, Li F, Zhou Q, Ohno S, Chen R, Kijlstra A, Rosenbaum JT, Yang P. Identification of a susceptibility locus in STAT4 for Behçet's disease in Han Chinese in a genome-wide association study. Arthritis Rheum. 2012;64(12):4104–13. https://doi.org/10.1002/art.37708.
40. Qin J, Li L, Zhang D, Yu H, Tan H, Zhang J, Deng B, Kijlstra A, Yang P. Analysis of receptor tyrosine kinase genetics identifies two novel risk loci in GAS6 and PROS1 in Behçet's disease. Sci Rep. 2016;6(1):26662. https://doi.org/10.1038/srep26662.
41. Qin X, Xu J, Wu Z, Sun F, Chen H, Zheng W, Li S, Li P, Chen S, Shen M, Zhang W, You X, Wu Q, Zhang F, Li Y z. Association study of rs924080 and rs11209032 polymorphisms of IL23R-IL12RB2 in a Northern Chinese Han population with Behcet's disease. Hum Immunol. 2016;77(12):1284–90. https://doi.org/10.1016/j.humimm.2016.09.006.
42. Xavier JM, Shahram F, Sousa I, Davatchi F, Matos M, Abdollahi BS, Sobral J, Nadji A, Oliveira M, Ghaderibarim F, Shafiee NM, Mojarad N. FUT2: filling the gap between genes and environment in Behçet's disease. Ann Rheum Dis. 2015;74:618–24. https://doi.org/10.1136/annrheumdis-2013-204475.
43. Alipour S, Nouri M, Sakhinia E, et al. Epigenetic alterations in chronic disease focusing on Behçet's disease: review. Biomed Pharmacother. 2017;91:526–33. https://doi.org/10.1016/j.biopha.2017.04.106.
44. Chang JT, Shevach EM, Segal BM. Regulation of interleukin (IL)-12 receptor β2 subunit expression by endogenous IL-12: a critical step in the differentiation of pathogenic autoreactive T cells. J Exp Med. 1999;189(6):969–78. https://doi.org/10.1084/jem.189.6.969.
45. Gheita TA, Gamal SM, Shaker I, El Fishawy HS, El Sisi R, Shaker OG, Kenawy SA. Clinical significance of serum interleukin-23 and A/G gene (rs17375018) polymorphism in Behçets disease: relation to neuro-Behçet, uveitis and disease activity. Joint Bone Spine. 2015;82(3):213–5. https://doi.org/10.1016/j.jbspin.2014.10.008.
46. Steinman L. Mixed reesults with modulation of T H-17 cells in human autoimmune diseases. Nat Immunol. 2010;11:41–4. Nature Publishing Group. https://doi.org/10.1038/ni.1803.
47. Çavuş F, Ulusoy C, Örçen A, Gül A, Tüzün E, Vural B. Increased IL-23 receptor, TNF-α and IL-6 expression in individuals with the IL23R-IL12RB2 locus polymorphism. Immunol Lett. 2014;160(1):96–8. https://doi.org/10.1016/j.imlet.2014.01.001.
48. Morton LT, Situnayake D, Wallace GR. Genetics of Behçet's disease. Curr Opin Rheumatol. 2016;28(1):39–44. https://doi.org/10.1097/BOR.0000000000000234.
49. Wolf SF, Sieburth D, Sypek J. Interleukin 12: a key modulator of immune function. Stem Cells. 1994;12(2):154–68. https://doi.org/10.1002/stem.5530120203.
50. Kappen JH, Medina-Gomez C, Van Hagen PM, Stolk L, Estrada K, Rivadeneira F, Uitterlinden AG, Stanford MR, Ben-Chetrit E, Wallace GR, Soylu M, Van Laar JAM. Genome-wide association study in an admixed case series reveals IL12A as a new candidate in Behçet disease. PLoS One. 2015;10(3):e0119085. https://doi.org/10.1371/journal.pone.0119085.
51. Sousa I, Shahram F, Francisco D, Davatchi F, Abdollahi BS, Ghaderibarmi F, Nadji A, Shafiee NM, Xavier JM, Oliveira SA. Brief report: association of CCR1, KLRC4, IL12A-AS1, STAT4, and ERAP1 with Behçet's disease in Iranians. Arthr Rheumatol (Hoboken, NJ). 2015;67(10):2742–8. https://doi.org/10.1002/art.39240.

52. Ortiz-Fernández L, Carmona FD, Montes-Cano MA, García-Lozano JR, Conde-Jaldón M, Ortego-Centeno N, Castillo MJ, Espinosa G, Graña-Gil G, Sánchez-Bursón J, Juliá MR, Solans R, Blanco R, Barnosi-Marín AC, De la Torre RG, Fanlo P, Rodríguez-Carballeira M, Rodríguez-Rodríguez L, Camps T, Castañeda S, Alegre-Sancho JJ, Martín J, González-Escribano MF. Genetic analysis with the immunochip platform in Behçet disease. Identification of residues associated in the HLA class I region and new susceptibility loci. PLoS One. 2016;11(8):e0161305. https://doi.org/10.1371/journal.pone.0161305.
53. Yazici H. Behçet's syndrome in the 2000s: "where is the wisdom we have lost in knowledge?". Clin Exp Rheumatol. 2016;34:23–5.
54. Wallace GR. HLA-B*51 the primary risk in Behçet disease. Proc Natl Acad Sci U S A. 2014;111(24):8706–7. https://doi.org/10.1073/pnas.1407307111.
55. Goswami R, Kaplan MH. STAT transcription factors in T cell control of health and disease. In: International review of cell and molecular biology, vol. 31: Academic Press; 2017. p. 123–80. https://doi.org/10.1016/bs.ircmb.2016.09.012.
56. Kirino Y, Bertsias G, Ishigatsubo Y, Mizuki N, Tugal-Tutkun I, Seyahi E, Ozyazgan Y, Sacli FS, Erer B, Inoko H, Emrence Z, Cakar A, Abaci N, Ustek D, Satorius C, Ueda A, Takeno M, Kim Y, Wood GM, Ombrello MJ, Meguro A, Gül A, Remmers EF, Kastner DL. Genome-wide association analysis identifies new susceptibility loci for Behçet's disease and epistasis between HLA-B*51 and ERAP1. Nat Genet. 2013;45(2):202–7. https://doi.org/10.1038/ng.2520.
57. Hou S, Xiao X, Li F, Jiang Z, Kijlstr A, Yang P. Two-stage association study in Chinese Han identifies two independent associations in CCR1/CCR3 locus as candidate for Behçet's disease susceptibility. Hum Genet. 2012;131(12):1841–50. https://doi.org/10.1007/s00439 012 1200 4.
58. Boone DL, Turer EE, Lee EG, Ahmad RC, Wheeler MT, Tsui C, Hurley P, Chien M, Chai S, Hitotsumatsu O, McNally E, Pickart C, Ma A. The ubiquitin-modifying enzyme A20 is required for termination of Toll-like receptor responses. Nat Immunol. 2004;5(10):1052–60. https://doi.org/10.1038/ni1110.
59. Zhou Q, Wang H, Schwartz DM, Stoffels M, Hwan Park Y, Zhang Y, Yang D, Demirkaya E, Takeuchi M, Tsai WL, Lyons JJ, Yu X, Ouyang C, Chen C, Chin DT, Zaal K, Chandrasekharappa SC, Hanson EP, Yu Z, Mullikin JC, Hasni SA, Wertz IE, Ombrello AK, Stone DL, Hoffmann P, Jones A, Barham BK, Leavis HL, Van Royen-Kerkof A, Sibley C, Batu ED, Gül A, Siegel RM, Boehm M, Milner JD, Ozen S, Gadina M, Chae J, Laxer RM, Kastner DL, Aksentijevich I. Loss-of-function mutations in TNFAIP3 leading to A20 haploinsufficiency cause an early-onset autoinflammatory disease. Nat Genet. 2015;48(1):67–73. https://doi.org/10.1038/ng.3459.
60. Li H, Liu Q, Hou S, Du L, Zhou Q, Zhou Y, Kijlstra A, Li Z, Yang P. TNFAIP3 gene polymorphisms confer risk for Behcet's disease in a Chinese Han population. Hum Genet. 2013;132(3):293–300. https://doi.org/10.1007/s00439-012-1250-7.
61. Ortiz-Fernandez L, García-Lozano JR, Montes-Cano MA, Conde-Jaldon M, Ortego-Centeno N, Garcia-Hernandez FJ, Espinosa G, Graña-Gil G, Sánchez-Bursón J, Juliá MR, Blanco R, Barnosi-Marín AC, Solans R, Fanlo P, Rodríguez Carballeira M, Camps T, Castañeda S, Martín J, González-Escribano MF. Lack of association of TNFAIP3 and JAK1 with Behçet's disease in the European population. Clin Exp Rheumatol. 2015;33(6, supl.94):S36–9. Retrieved from http://www.ncbi.nlm.nih.gov/pubmed/26005883
62. Di Marzio P, Dai WW, Franchin G, Chan AY, Symons M, Sherry B. Role of Rho family GTPases in CCR1- and CCR5-induced actin reorganization in macrophages. Biochem Biophys Res Commun. 2005;331(4):909–16. https://doi.org/10.1016/J.BBRC.2005.04.015.
63. Karasneh J, Gu A, Ollier WE, Silman AJ, Worthington J. Whole-genome screening for susceptibility genes in multicase families with Behçet's disease. Arthr Rheumatol. 2005;52(6):1836–42. https://doi.org/10.1002/art.21060.
64. Ciucci T, Bosselut R. Gimap and T cells: a matter of life or death. Eur J Immunol. 2014;44(2):348–51. https://doi.org/10.1002/eji.201344375.

65. Ferrer-Admetlla A, Sikora M, Laayouni H, Esteve A, Roubinet F, Blancher A, Calafell F, Bertranpetit J, Casals F. A natural history of FUT2 polymorphism in humans. Mol Biol Evol. 2009;26(9):1993–2003. https://doi.org/10.1093/molbev/msp108.
66. Archer DA, Graham ME, Burgoyne RD. Complexin regulates the closure of the fusion pore during regulated vesicle exocytosis. J Biol Chem. 2002;277(21):18249–52. https://doi.org/10.1074/jbc.C200166200.
67. Uğurel E, Şehitoğlu E, Tüzün E, Kürtüncü M, Çoban A, Vural B. Increased complexin-1 and decreased miR-185 expression levels in Behçet's disease with and without neurological involvement. Neurol Sci. 2016;37(3):411–6. https://doi.org/10.1007/s10072-015-2419-3.
68. Fei Y, Webb R, Cobb BL, Direskeneli H, Saruhan-Direskeneli G, Sawalha AH. Identification of novel genetic susceptibility loci for Behçet's disease using a genome-wide association study. Arthritis Res Ther. 2009;11(3):R66. https://doi.org/10.1186/ar2695.
69. Hou S, Yang P, Du L, Zhou H, Lin X, Liu X, Kijlstra A. SUMO4 gene polymorphisms in Chinese Han patients with Behcet's disease. Clin Immunol. 2008;129(1):170–5. https://doi.org/10.1016/j.clim.2008.06.006.
70. Kim J, Im CH, Kang EH, Lee EY, Lee YJ, Park KS, Song YW. Mannose-binding lectin gene-2 polymorphisms and serum mannose-binding lectin levels in Behçet's disease. Clin Exp Rheumatol. 2009;27(2):S13–7.
71. Boiardi L, Salvarani C, Casali B, Olivieri I, Ciancio G, Cantini F, Salvi F, Malatesta R, Govoni M, Trotta F, Filippini D, Paolazzi G, Nicoli D, Farnetti E, Macchioni L. Intercellular adhesion molecule-1 gene polymorphisms in Behçet's disease. J Rheumatol. 2001;28(6):1283–7. Retrieved from http://www.ncbi.nlm.nih.gov/pubmed/11409120.
72. Pulsatelli L. Vascular endothelial growth factor gene polymorphisms in Behçet's disease. J Rheumatol. 2004;31(9):1785–9.
73. Burillo-Sanz S, Montes-Cano M-A, García-Lozano J-R, Ortiz-Fernández L, Ortego-Centeno N, García-Hernández F-J, Espinosa G, Graña-Gil G, Sánchez-Bursón J, Juliá MR, Solans R, Blanco R, Barnosi-Marín AC, De la Torre RG, Fanlo P, Rodríguez-Carballeira M, Rodríguez-Rodríguez L, Camps T, Castañeda S, Alegre-Sancho JJ, Martín J, González-Escribano MF. Mutational profile of rare variants in inflammasome-related genes in Behçet disease: a next generation sequencing approach. Sci Rep. 2017;7(1):8453. https://doi.org/10.1038/s41598-017-09164-7.
74. Wu Z, Zhang S, Li J, Chen S, Li P, Sun F, Wen X, Zheng W, Zhang F, Li Y. Association between MEFV mutations M694V and M680I and Behçet's disease: a meta-analysis. PLoS One. 2015;10(7):e0132704. https://doi.org/10.1371/journal.pone.0132704.
75. Okabe K, Izumi Y, Miyashita T, Irino K, Kawahara C, Jiuchi Y, Nonaka F, Eguchi K, Kawakami A, Migita K. An usual case of vasculo- and neuro-Behçet's disease with MEFV mutations. JPN J Clin Immunol. 2014;37(1):61–7. https://doi.org/10.2177/jsci.37.61.
76. Yüksel Ş, Kucukazman SO, Karataş GS, Ozturk MA, Prombhul S, Hirankarn N. Methylation status of Alu and LINE-1 interspersed repetitive sequences in Behcet's disease patients. Biomed Res Int. 2016;2016:1–9. https://doi.org/10.1155/2016/1393089.
77. Alipour S, Nouri M, Sakhinia E, Samadi N, Roshanravan N, Ghavami A, Khabbazi A. Epigenetic alterations in chronic disease focusing on Behçet's disease: review. Biomed Pharmacother. 2017;91:526–33. https://doi.org/10.1016/j.biopha.2017.04.106.

Immunology of Neuro-Behcet's Disease (NBD)

Güher Saruhan-Direskeneli and Haner Direskeneli

Behcet's disease (BD) is a systemic inflammatory disorder with a diverse spectrum of clinical manifestations including mucocutaneous, ocular, vascular, gastrointestinal, musculoskeletal, and central nervous system (CNS) involvement. Neurological involvement of BD is relatively rare. According to an epidemiological survey in Japan in 1972, neuro-Behcet's disease (NBD) accounts for 13.7% of 1876 cases of BD [1]. There are reports of 5–30% of cases with NBD in the long-term follow-up of BD patients.

A complex genetic background leading to a pro-inflammatory, innate immune system-derived activation perpetuated by adaptive immune responses against environmental antigens and/or auto-antigens is accepted as the main pathogenic mechanism in BD. Immunological mechanisms underlying the inflammation have been investigated for infectious agents or inflammatory stimuli and the innate and adaptive responses of the patients. In NBD, the specific features of the CNS, but accessibility of the cerebrospinal fluid (CSF) as the local tissue specimen, provided a challenge for immunological studies compared to BD.

G. Saruhan-Direskeneli (✉)
Department of Physiology, Istanbul University, Istanbul Medical Faculty, Istanbul, Turkey
e-mail: gsaruhan@istanbul.edu.tr

H. Direskeneli
Department of Internal Medicine, Division of Rheumatology, Marmara University, School of Medicine Hospital, Istanbul, Turkey

E. Tüzün, M. Kürtüncü (eds.), *Neuro-Behçet's Disease*,
https://doi.org/10.1007/978-3-030-55273-2_2

Immune Response in the CSF

Cellularity of CSF

As a local tissue response, the increased cellularity has been considered as an indication of immune activation, and differential evaluation of the cell content pointed at the etiological agents of the stimulus. In an extended series of parenchymal NBD (p-NBD), cellular examination of CSF has revealed normal findings as well as pleocytosis in about half of the cases. If pleocytosis was present, the CSF cell count was in the range of 0–400/μl/mm^3 (median, 30), and in about half of these patients, neutrophilic predominance or both neutrophils and lymphocytes were detected. In the other half of the patients' CSF, higher percentages of mononuclear cells and lower percentages of activated lymphocytes were seen. The pleocytosis in NBD represented perivascular inflammatory process in the CNS [2, 3].

Considering the neutrophilic pleocytosis in NBD, involvement of the innate immune mechanisms in disease pathogenesis is implicated. Typical BD lesions such as pustular folliculitis, pathergy reaction, and hypopyon have significant neutrophil infiltrates. Therefore, neutrophil function and activation status have been extensively investigated. Conflicting reports of increased, normal, or decreased superoxide production, phagocytosis, chemotaxis, and neutrophil-endothelial adhesion are available. These might reflect differing clinical activity, the discordance between "in vitro" and "in vivo" state of affairs, drug effects, or general methodological problems [4–6].

Two clinical categories of NBD are differentiated: p-NBD (intra-axial NBD, primary-NBD, or "NBD"), which is caused by parenchymal pathology, accounts for the majority of NBD. p-NBD is characterized by diffuse brainstem, cerebral, optic, and spinal cord symptoms. On the other hand, nonparenchymal-NBD (np-NBD, vasculo-NBD, secondary-NBD, or extra-axial NBD) [7–9], which is usually caused by occlusion or hemorrhage of the main vascular structures or aneurysm in the CNS, is relatively rare. The frequency of this subtype is reported to be 10% to 20% of all NBD patients from Turkey [2, 10]. Cerebral venous thrombosis, pseudotumor-like intracranial hypertension, and acute meningeal syndrome are common forms of np-NBD. Hirohata et al. have proposed a clinical-orientated and simple classification in which p-NBD is further classified into acute and chronic progressive types, depending on its clinical course [11].

Immune Mediators as Indications of Activity in CSF

In an early study, serial CSF examinations in a patient with BD involving CNS have been performed. CSF indexes (ratios to serum levels) of immunoglobulins (Ig) and the third (C3) and the fourth (C4) components of complement were measured, and immune complexes and lymphocyte subsets were quantified in CSF and peripheral blood (PB) before and after immunosuppressive treatment. During active encephalitis, intrathecal production of IgM and, to a lesser degree, IgG and IgA, presence of immune complexes, intrathecal C3 production, and elevated C3 and C4 concentrations in CSF but not in serum were detected. An increased CSF $CD8^+$ T cell percentage, in combination with slightly increased PB $CD3^+$ and $CD8^+$ T cell subsets, was shown. Further data have also demonstrated an increase of IgG index in 73% of patients. After effective immunosuppressive treatment, humoral and cellular CSF values were normal. Intrathecally produced Ig, immune complexes, and C3 as well as $CD8^+$ T cells are likely to participate in the development of "Behcet encephalitis" [2, 12].

Oligoclonal Bands (OCB)

OCB of Ig in the CSF provides an evidence for a humoral response. As a possible indication of intrathecal antibody production, OCB of IgG, A, and M were investigated firstly in 13 BD patients with CNS involvement and in 40 neurologic controls. Oligoclonal IgA and IgM bands were detected mainly in CSF samples from patients with active NBD and were documented to disappear when neurologic manifestations remit. High Ig index values were detected in both active and quiescent diseases. Based on these findings, CSF oligoclonal IgA and IgM were recommended in monitoring CNS disease activity in NBD [13].

A retrospective study in CSF and serum samples from 12 patients with BD with neurological involvement found local synthesis of OCB of IgG using isoelectric focusing and immunoblotting in 8% at some stage in their disease. When assessed with an albumin ratio of CSF to serum, 42% of patients with BD exhibited an abnormal barrier function at the same time. Serial CSF analysis showed that clinical relapses were associated with worsening barrier function and in some patients the development of local oligoclonal IgG synthesis; conversely, steroid treatment led to a statistically significant improvement in barrier function and in two patients a loss of oligoclonal IgG bands. OCB was less frequent in NBD than sarcoidosis (51%) and systemic lupus erythematosus (SLE) (25%) in this series [14].

In more recent and extended series of paired CSF and serum samples from 74 p-NBD patients, 22 patients with cerebral venous sinus thrombosis (CVST) and BD, 18 with primary headache disorders not directly associated with BD, and 7 with cerebrovascular event, intrathecal production of IgG (pattern 2 or 3) was considered as positive. Totally, 10.8% of patients had OCB in the CSF only (pattern 2). These positive cases had p-NBD. All other groups were negative. In the differential diagnosis of chronic neurological disorders, the absence of bands was suggested as a useful diagnostic clue to differentiate NBD from diseases such as multiple sclerosis (MS) [15].

Other Markers in the CSF

The presence of myelin basic protein (MBP) from the CNS myelin has been used as an indication of myelin damage in the CNS. High levels of MBP were detected in the CSF from four patients with NBD and none of the screened other neurological diseases. A good correlation between clinical states and the levels of MBP was also observed in this small study, albeit the presence of MBP is frequent in many disease states [16].

Similarly, another nonspecific marker of immune response, beta$_2$-microglobulin (β_2MG), has been measured in paired serum and CSF specimens from patients with NBD and non-inflammatory neurological diseases (NIND) along with albumin. Albumin index (Q albumin as an indicator of blood-brain barrier function) and β_2MG index (an indicator of intrathecal β_2MG synthesis) were significantly elevated in patients with NBD compared with the control patients. In about half of the patients with NBD, CSF β_2MG and Q albumin were significantly decreased when the CNS manifestations were improved by successful treatment. The elevation of CSF β_2MG with the barrier dysfunction has implicated the transudation of serum β_2MG as well as the increased intrathecal synthesis presumably by infiltrating lymphocytes, reflecting CNS disease activity in NBD [17].

Cytokines

Cytokines and chemokines as the humoral mediators of the immune response have also been investigated in paired CSF and serum specimens of patients. As an inflammatory cytokine with wide effects, interleukin (IL)-6 has been evaluated in NBD firstly. IL-6 and anti-cardiolipin antibodies, especially of the IgM isotype, were highly elevated in the CSF of only five patients with NBD. As levels of both dropped after disease activity subsided, serial measurements of IgM anti-cardiolipin antibodies and IL-6 in the CSF were suggested as useful in NBD [18]. Emphasizing the role of this cytokine, functional IL-6 activity in patients with progressive NBD compared with active BD but lacking progressive CNS disease has shown marked elevation of IL-6 activity in the CSF, but not in the serum. There was no significant correlation of CSF IL-6 activity with serum IL-6 activity, CSF cell counts, CSF total protein levels, or the CSF/serum albumin quotient. A role of persistent chronic CNS inflammation has been suggested to play an important role in the pathogenesis of progressive neuropsychiatric manifestations in BD [19].

Following this observation, a more extensive group of 68 BD patients with acute or chronic progressive p-NBD, dural sinus thrombosis, ischemic stroke, or headache were compared with MS, subacute sclerosing panencephalitis, and NIND for CSF IL-6 levels. CSF but not serum samples of NBD patients with acute p-NBD displayed significantly increased IL-6 levels as compared to other groups. Chronic progressive p-NBD patients also showed a less prominent increase in CSF IL-6. CSF IL-6 has been demonstrated as a marker of disease activity and long-term outcome for p-NBD along with CSF cell count and protein levels. CSF IL-6 could be used in chronic progressive patients who have normal CSF cell or protein levels to detect disease activity [20].

A comparison of cytokines (IL-10, IL-12p70, IL-17A) and chemokines (CCL2, CXCL10, CXCL8 (IL-8)) included patients with NBD, MS, infectious and/or inflammatory neurological diseases (IND), and other NIND. In the CSF, CXCL10 levels were significantly higher in NBD and IND than NIND and MS, whereas CXCL8 was increased in NBD compared to NIND. IL-12 was elevated in the CSF of IND compared to NBD and NIND and also in the CSF of MS compared to NIND. By these results NBD resembled nonspecific inflammations such as neuro-infections compared to autoimmune disorders such as MS. An unknown infection as a trigger of a vasculitic process in the CNS has been hypothesized [21]. When IL-6, IL-8, IL-10, TNF-α, and IFN-γ were measured, patients with viral meningitis had significantly higher levels of all investigated cytokines except for IFN-γ in comparison with patients with p-NBD, headache attributed to BD (HaBD) and HC. Higher CSF IL-6 was confirmed in patients with p-NBD. No analogy between CSF cytokine profiles of patients with p-NBD and viral meningitis, but a crucial role for IL-6 in immunopathogenesis of NBD has been repeatedly reported [22].

Another value of cytokine measurements as biomarkers has been the evaluation of their relationship with disease activity. CSF total cell and polymorphonuclear leukocyte counts were significantly lower in the patients with chronic progressive NBD than in those with acute NBD. The CSF levels of IL-6 and IL-8 were markedly elevated in the NBD patients compared with those measured in the control patients with NIND with no significant differences between the patients with acute and chronic progressive NBD. There were no significant differences in the CSF levels of IL-1β and TNF-α among the control, acute NBD, and chronic progressive NBD patients. The CSF levels of IL-6 and IL-8 were significantly decreased following successful treatment in both acute NB and chronic progressive NB patients, whereas the CSF levels of IL-1β and TNF-α were not changed significantly. The mechanisms underlying the elevation of CSF IL-6 and IL-8 might be different in patients with acute NB and those with chronic progressive NB [11]. Interestingly, serum IL-23 levels were significantly higher in patients with NBD among a BD patient group. The precise role of IL-23 remains elusive. However, IL-23 receptor is upregulated in the presence of IL-6 [23]. Acute p-NBD typically features acute and transient symptoms such as fever and hemiparesis accompanied by inflammatory features including elevated cell count in the CSF, while chronic progressive p-NBD is characterized by ataxia, dementia, incontinence, and brainstem atrophy.

Several other cytokines with pro-inflammatory activity have also been screened in NBD samples mostly in single studies. Macrophage migration inhibitory factor (MIF), released by activated T lymphocytes and macrophages, is required for antigen- and mitogen-driven T cell activation and stimulates macrophages to release cytokines and nitric oxide. The determination of MIF in the CSF of patients with NBD and NIND revealed that in NBD patients, the concentration of MIF was significantly elevated compared with control samples and correlated well with cell counts in these samples. With these findings, a role for MIF in immune-mediated diseases of the CNS has been introduced, but not further followed in NBD [24].

When IL-15 was measured in serum and CSF samples of patients with BD, IND, and NIND, active BD patients have significantly higher serum IL-15 levels compared with BD in remission and HC. Similar serum IL-15 levels were found in active NBD and IND groups. Elevated levels of IL-15 were observed in CSF samples from NBD patients. IL-15 was involved in BD inflammatory process, particularly in vasculitis foci, as an elevated CSF/serum IL-15 ratio characterizes vascular cerebral lesions [25].

IL-33 is released from damaged tissues or necrotic cells and acts as an alarmin in the host defense against pathogens. In NBD patients, higher IL-33 levels in CSF and higher IL-33 mRNA transcripts in CSF mononuclear cells were shown compared with HaBD and NIND patients. IL-33 levels were also remarkably higher in CSF than in serum. Relatedly, NF-κB DNA binding activity and expression of IP-10 and MCP-1 were high in NBD patients compared to NIND and HaBD patients. Considering the hypothesis of an infectious trigger in BD pathogenesis, IL-33, together with MCP-1 and IP-10, has been suggested as critically required for fighting the hypothetical pathogen, as IL-33 may act as a potential regulator of innate immune responses in the CNS [26].

As inaugural clinical symptoms for both NBD and MS might be similar, possible differences in the expression profiles in the blood and the CSF of MS and NBD patients at early stages were screened for discriminative markers. Cytokines and transcription factors related to Th1, Th2, Th17, and T regulatory populations were evaluated simultaneously in PB and CSF, from 40 patients presenting a first episode of clinical features related to CNS inflammation and 22 controls with NIND enrolled mainly for severe headache. At the follow-up, 21 patients were diagnosed with relapsing-remitting MS (RRMS) and 19 had NBD. In initial blood samples, T-bet expression was significantly increased in NBD patients only, while IFN-γ was elevated in patients who evolved into RRMS or NBD. IL-17A, GATA-3, and IL-4 were significantly lower in RRMS patients than in the NBD group. In the initial CSF samples, ROR-γt, IL-17A, and IFN-γ were significantly elevated in patients compared to controls. Most strikingly, a significant increase of CSF IL-10 was observed in NBD patients only, and CSF IL-10 was proposed as a predictive marker to discriminate between these two disorders [27]. An inflammatory profiling (MMP9, TNFα, IL-6, CXCL13, CXCL10, CXCL8, IFN-γ, IL-10, IL-17, IL-23, and others) between NBD with MS-like features and MS in the CSF and serum samples revealed that only MMP9 was increased in NBD serum compared to MS and decreased in CSF. Furthermore, NBD analysis of circulating natural killer $CD56^{DIM}$ subset suggests their potential involvement in increased MMP9 production [28].

A mainly B cell-related mediator, B cell-activating factor of the tumor necrosis factor family (BAFF) and its receptors' expression were also higher in CSF cells of NBD and MS patients compared to NIND patients. Serum BAFF levels were higher than CSF levels in NBD patients [29]. The increased levels of CSF BAFF in NBD patients were not significantly different from patients with epidemic aseptic meningitis (AM) and MS, but from healthy controls (HC). CSF BAFF levels were significantly increased in patients with a slowly progressive course compared with those with an acute course and did not correlate with serum BAFF levels, CSF cell counts, or CSF IL-6 levels. BAFF production within the CNS is suggested to be associated with a progressive course of NBD [30].

Vascular endothelial growth factor (VEGF) stimulates mainly angiogenesis and has pro-inflammatory effects. Significantly increased CSF VEGF levels in NBD and MS patients were detected compared to NIND patients. An association between CSF NBD-VEGF and leukocyte count and CSF CD4 cells as well as VEGF mRNA was demonstrated without further data on its effect [31].

Candidate Antigens of Possible Autoimmune Response in NBD

BD does not have the classical clinical features of autoimmunity such as female dominance or association with other autoimmune diseases such as in Sjogren's syndrome [32, 33]. However, BD has various aspects that deserve to be considered as "autoimmune." Classical, systemic disease-associated autoantibodies are not observed; however, various antibodies against specific antigens such as α-enolase, α-tropomyosin, kinectin, selenium-binding protein, esterase D, and carboxy-terminal subunit of Sip1 have been shown in BD sera [34]. Most of these autoantibodies were demonstrated in uveitis patients, and their role in disease pathogenesis is not yet clear. Autoimmune and dysimmune inflammatory mechanisms on a genetically susceptible background are implicated in the etiology of BD.

Heat Shock Protein (HSP) and Related Responses

For an autoimmune etiology, various candidate auto-antigens have also been investigated in BD patients for T cell reactivity. Among the antigens, human HSP60 has been the one most extensively studied [35].

Systemic immune reactivity to 65-kD mycobacterial HSP65 (m-HSP65) has been shown in BD [36]. T cell response to the 65-kDa HSP in patients with BD was mapped to four HSP peptides: 111–125, 154–172, 311–325, and 219–233. Higher frequency of short-term T cell lines (STCL) in BD did not correspond to any HLA-DR. Being a susceptibility marker for BD, HLA-B51 was not a restricting element for these T cells. Similarly, higher lymphoproliferative responses were stimulated by the human HSP sequence-derived peptides compared with the corresponding mycobacterial peptides supporting the involvement of 65-kDa HSP in the pathogenesis of BD. Whereas the high microbial load and associated stress proteins found in oral ulceration of BD may initiate an immune response to these conserved epitopes, expression of autoreactive T cell clones might be stimulated by immunodominant T cell epitopes of endogenous HSP which may induce immunopathologic changes [37]. HSP65 derived from *Streptococcus sanguinis* was also proposed as a triggering factor based on its homology with human HSP60. Moreover, four immunodominant epitopes of HSP60 induced T as well as B cell responses in studies from UK, Japan, and Turkey. Th1 type, pro-inflammatory cytokine response to HSP60-derived peptide 336–51 with IFN-γ, IL-12, and TNF-α production is demonstrated. HSP60 peptide 336–51 also upregulated Txk, a tyrosine kinase expressed in Th1 cells and effected IFN-γ gene transcription [38]. B cell epitopes of the m-HSP65 were mapped in the sera from patients with BD. Significant increases in IgA and IgG antibodies to peptides 111–125, 154–172, and 311–326 and homologous peptides from the human mitochondrial HSP60 136–150 and 336–351 were detected in the sera from BD. Antibodies to each of the peptides represented a small proportion of the total B cell epitope repertoire elicited by the 65 or 60 kD HSP. As the peptides induced experimental uveitis in Lewis rats, similar to the principal manifestation of BD, the response to them has implicated the involvement in the immunopathogenesis of BD [39]. However, none of the autoantigens identified so far which share common epitopes with bacterial HSP65 has a high prevalence in the sera from BD.

Recently, immunoreactivity against filamentous neuronal processes in the mouse brain, retina, and scrotal skin was detected in great majority of BD patients' sera. This reactivity was evident in the sera of patients with (75%) and without (64%) neurological involvement. Neurofilament medium (NF-M) was identified as the probable antigen. Surprisingly, homology between the human NF-M and m-HSP65 corresponding to peptides 111–126, 213–232, and 304–363 has been demonstrated, and sera immunoreactive against NF-M cross-reacted with bacterial HSP65. With these findings, molecular mimicry of bacterial HSP65 with NF-M has also been implicated in the autoimmunity of BD [40].

Based on the reported importance of anti-HSP immune response in BD, NBD was also evaluated for the local response. The comparison of the intrathecal antibody production to m-HSP65 in the CSF of p-NBD patients with NIND and MS revealed significantly higher CSF IgG responses in p-NBD patients, with an anti-m-HSP65 positivity rate of 48%. CSF anti-m-HSP65 IgG ratios of CSF to serum correlated with the duration of BD, but not with the duration of neurological involvement. Serum IgM and IgA responses were elevated in patients with np-NBD with intracranial hypertension, suggesting a different type of involvement than p-NBD. With these results, an increased intrathecal humoral response to m-HSP65 in the CSF of p-NBD patients was also confirmed [41]. Another small HSP with increased tissue expression in demyelination has also been a target for antibodies in NBD. Serum and CSF IgG antibody responses to αB-crystallin were significantly elevated in NBD patients compared to NIND patients. Serum IgM anti-αB-crystallin antibody titers were also higher in NBD as in MS [42].

Neuronal and Other Proteins

When antibodies against neuronal antigens were screened by immunochemistry on hippocampal and cerebellar molecular layers as well as hippocampal neuron cultures, reactivity was detected in sera and/or CSF of 13 of 20 NBD and 6 of 20 BD patients but not in MS or headache controls. Screening with a protein macroarray led to the identification of stress-induced phosphoprotein-1 (STIP-1) and mitochondrial carrier homolog 1 (Mtch1), an apoptosis-related protein, as potential autoantigens. High-titer STIP-1-antibodies were detected in six NBD patients' sera but not in controls [43]. Serum Mtch1 antibodies in 68 of 144 BD patients with or without neurological involvement and in 4 of 168 controls corresponded to a sensitivity of 47.2% and a specificity of 97.6% for these antibodies. Mtch1 antibody-positive NBD patients had more attacks, increased disability, and lower serum nucleosome levels. Mtch1 antibody might be involved in pathogenic mechanisms of NBD rather than being a coincidental by-product of auto-inflammation [44].

Recently, in two Japanese patients who had been diagnosed with probable NBD, myelin oligodendrocyte glycoprotein (MOG) antibody-associated relapsing encephalitis is documented. So, the presence of anti-MOG antibodies has to be differentiated from probable NBD because both disorders can present with brainstem or cerebral lesions, CSF pleocytosis, and elevated levels of CSF IL-6 and respond to steroid treatment.

Cellular Immune Response in NBD

There are very scarce data on the pathological findings at the tissue level in NBD. In an early study, the presence of scattered foci of necrosis, demyelination, and scars throughout the CNS, frequently in close proximity to small arterioles or venules infiltrated with inflammatory cells, was described [45]. Immunohistological examination of biopsied or autopsied brain tissues from three patients with acute NBD, chronic progressive NBD, and NBD in a long-term remission was more informative and demonstrated the cellular response. Histopathology in acute NBD lesion revealed infiltration of mononuclear cells around small vessels, consisting of $CD45RO^+$ T lymphocytes and $CD68^+$ monocytes with few $CD20^+$ B lymphocytes. Most neurons were undergoing apoptosis in the inflammatory lesion. In chronic progressive NBD, similar histopathological changes were noted in pons, cerebellum, medulla, internal capsule, and midbrain with modest degree of mononuclear cell infiltration and scattered foci of neuronal apoptosis with formation of a few binucleated neurons. The prominent feature of NBD in a long-term remission was atrophy of basal pons with formation of cystic or moth-eaten lesions, consisting of isomorphic gliosis with viable neurons and scattered foci of perivascular cuffing of T lymphocytes and monocytes. The common features throughout the courses of NBD were perivascular cuffing of T lymphocytes and monocytes, irrespective of the clinical phenotypes. Based on these observations, soluble factors produced by infiltrating cells, including IL-6, have been accounted for the induction of apoptosis of neurons in NBD [46].

As NBD can present with diverse clinical and pathological manifestations, few cases presenting with neurological symptoms preceding other systemic features are difficult to be diagnosed. In a case report from Japan, brain biopsy has been suggested as a useful method to differentiate a mass with perivascular infiltration of neutrophils in the meninges with intact brain parenchyma as NBD from a tumor [47].

In BD, evidence for the presence and activation of T cells both in the PB and tissue specimens is observed. Both $CD4^+$ and $CD8^+$ T cells producing pro-inflammatory cytokines IL-2 and IFN-γ are shown to be increased in the PB and correlated with disease activity [48]. IL-12, which drives Th1 response in naive T cells, was elevated in patient sera, in correlation with Th1 lymphocytes [49–51]. Th17 subset producing IL-17A, IL-17F, IL-22, and TNF-α [52] and induced by IL-1, IL-6, and TGF-β to differentiate is suggested to be involved in organ-specific autoimmunity. Active BD is characterized by high levels of IL-6, IL-10, and IL-17 compared to remission [53]. Levels of IL-23 and IFN-γ were also elevated in BD patients with active uveitis [54]. IL-21 was shown to affect Th1 and Th17 differentiation and lead to a decrease of regulatory T cells [55]. In contrast, IL-27 inhibits the production of IL-1β, IL-6, and IL-23 but promotes IL-10 production by dendritic cells (DCs), which inhibit both Th1 and Th17 cell responses. Decrease in IL-27 expression has been shown in active BD with uveitis [56, 57].

In NBD patients, expression of lineage-specific transcription factors of Th cells, TBX2/RORC (Th1), RORC (Th17), and FOXP3 (Treg) were increased in the CSF compared to HaBD and NIND patients. EBI3 and Th2-associated GATA3 expressions were found to be decreased in NBD patients. Analysis of transcription factor ratios revealed an increase in the RORC/FOXP3 and TBX21/GATA3 ratios implicating activation of inflammatory T cell subpopulations and dysregulations of Th1, Th17, and Treg cells in CSF-NBD inflammation [58].

Recently, pathway analysis by mRNA profiling has been applied to BD at the cellular level. JAK/STAT signaling was identified as activated both in $CD14^+$ monocytes and $CD4^+$ lymphocytes in BD patients, suggesting an activation of both innate and adaptive immunity [59]. Interferon-γ, glucocorticoid receptor, and IL-6 signaling was also prominent among upregulated genes in $CD14^+$ monocytes. In another study, neuregulin signaling pathway molecules (epiregulin (EREG), amphiregulin (AREG), and neuregulin-1 (NRG1)) were overrepresented among the differentially expressed genes, suggesting a role of EGF/ErbB signaling pathway in BD pathogenesis [60]. Finally, in a recent study, Th1 and Th17 cells were again implicated, and clusters enriched in T and B cell activation with type I IFN, JAK/STAT, and TLR signaling pathways were demonstrated [61]. Genome-wide association study (GWAS) data were also used to analyze biochemical pathways and protein-protein interaction networks through the single-nucleotide polymorphisms detected in BD. Using Turkish and Japanese GWAS data, focal adhesion, MAPK signaling, TGF-β signaling, ECM-receptor interaction, complement/coagulation cascades, and proteasome pathways were implicated as shared pathways associated with the genetic risks of BD [62]. With a proteomics of PB, tyrosine-protein phosphatase non-receptor type 4, threonine synthase-like 2, and β-actin have been demonstrated as discriminators between the BD and non-BD groups and were suggested as possible biomarkers [63].

Other Mechanisms in BD

Inflammation, Coagulation, and Fibrinolytic Pathway Abnormalities

Vascular involvement in BD is predominantly venous in contrast to what is seen in other systemic vasculitides. Dural/sagittal sinus thrombosis is accepted to be a part of NBD, but possibly more associated with vascular pathogenic factors. In histopathological specimens, in addition to thrombi, inflammatory infiltrates in vessel walls point to a vasculitic process in BD [64]. No specific defect in the coagulation cascade has so far been demonstrated [65, 66]. However, both coagulation and fibrinolytic pathways seem to be activated with or without thrombosis. Increased levels of thrombin-antithrombin (TAT) complex and prothrombin fragment 1 + 2 support intravascular thrombin generation in these patients as a result of the activation of the coagulation cascade. Various procoagulant conditions associated with an increased risk of thrombosis such as deficiencies of protein C, protein S and antithrombin III, factor V Leiden, and prothrombin 20210A mutations may also contribute to the prothrombotic state of BD. Severalfold increases in the risk of thrombosis have been described in carriers of factor V Leiden and prothrombin gene mutations in patients with BD [67].

In a recent study [68], thrombin-catalyzed fibrin formation and fibrin susceptibility to plasmin-induced lysis were significantly impaired in BD patients. These findings were associated with increased plasma oxidative stress markers and with a marked carbonylation of fibrinogen. In the same study, neutrophils displayed an enhanced NADPH oxidase activity and increased reactive oxygen species production, which significantly correlated with fibrinogen carbonylation level and fibrinogen clotting ability. These data suggest an important link between innate immune response and thrombosis in BD patients with vascular involvement. The implication of these findings in NBD has not been evaluated separately.

Severity and Gender

One currently underexplored area is the gender differences in BD, a condition pronouncedly more severe among males [69]. Severe complications such as vascular, CNS, and pulmonary involvement as well as mortality are related to the male gender. However, serum levels of testosterone and estradiol are not different among the male patients. Estrogen is shown to protect against endotoxin-induced uveitis (EIU) in Lewis rats. Estrogen decreases E-selectin and IL-6 gene expressions through estrogen receptors in vascular endothelium. As another mechanism, fMLP-stimulated superoxide generation from neutrophils is decreased in vitro with estrogen incubation [70]. Through these mechanisms, estrogens might suppress the pro-inflammatory functions of vascular endothelium and neutrophils, explaining the milder clinical course in females. Alternatively, it was shown that testosterone augments the neutrophil functions, especially in male BD patients [71]. Incubation with testosterone causes elevated IL-12/IL-2 and decreased IL-10 levels in BD, suggesting that it activates a Th1 phenotype [72].

Conclusions

It might be too simplistic to describe BD as either an autoimmune or an auto-inflammatory disease. An environmental agent is possibly required to trigger the inflammation in BD. It might then be linked to innate immune abnormalities which predispose to early or more intense neutrophil and monocyte responses [73]. However, unlike classical auto-inflammatory disorders, an adaptive response is also sustained through bacterial persistence or autoantigen-activated dendritic, T or B cells.

CSF studies, mostly performed in p-NBD, suggest that both innate and Th1/Th17-type adaptive immune responses are present with increased pro-inflammatory (IL-6, TNF-α, and IL-1) and adaptive immune cytokine elevations (IL-15, IL-33). However, the local specificity of this inflammatory milieu is not clearly defined compared to infections or autoimmune CNS disorders. Similarly, autoantibody responses against both local (neurofilament medium) and systemic autoantigens (HSPs) are present in CSF of p-NBD patients; however, whether these responses are pathogenic or just the result of nonspecific tissue injury is not clear as antigen-antibody complexes or antigen-specific T-cell responses leading to tissue damage in NBD tissue specimens are not demonstrated.

One important issue that needs to be explained is the highly specific brainstem involvement in NBD (quite different from demyelinating disorders). As BD is accepted to have a vasculitic component, these infiltrations are suggested to have a venous venulitis pattern which still needs to be explained for immune pathogenesis for all vascular BD manifestations. Nonspecific neutrophil/monocyte infiltrations are present in vasa vasorum of venous structures in vascular BD patients, but without a clear immune explanation yet. Although arterial involvement similar to large-vessel vasculitides such as aneurysms or occlusions is also observed in BD, it is not intra-cranially present, excluding an arterial pathogenesis in NBD.

Clarification of all these local and systemic immune mechanisms might help to elucidate how both antimicrobial and immunosuppressant therapies seem to be effective in BD or NBD and might pave the way for more specific immune interventions.

References

1. Shimizu T. [Behcet;s syndrome - lists of clinical and epidemical statistics] Nihon Rinsho. 1974;32 Spec No:2093–2102.
2. Akman-Demir G, Serdaroglu P, Tasci B. Clinical patterns of neurological involvement in Behcet's disease: evaluation of 200 patients. Neuro Behcet Study Group Brain. 1999;122(Pt 11):2171–82.
3. Nakamura S, Takase S, Itahara K. Cytological examination of cerebrospinal fluid in eight patients with neuro-Behcet's disease. Tohoku J Exp Med. 1980;132:421–30. https://doi.org/10.1620/tjem.132.421.
4. Eksioglu-Demiralp E, Direskeneli H, Kibaroglu A, Yavuz S, Ergun T, Akoglu T. Neutrophil activation in Behcet's disease. Clin Exp Rheumatol. 2001;19:S19–24.
5. Greco A, et al. Behcet's disease: new insights into pathophysiology, clinical features and treatment options. Autoimmun Rev. 2018;17:567–75. https://doi.org/10.1016/j.autrev.2017.12.006.
6. Mege JL, et al. Overproduction of monocyte derived tumor necrosis factor alpha, interleukin (IL) 6, IL-8 and increased neutrophil superoxide generation in Behcet's disease. A comparative study with familial Mediterranean fever and healthy subjects. J Rheumatol. 1993;20:1544–9.
7. Al-Araji A, Kidd DP. Neuro-Behcet's disease: epidemiology, clinical characteristics, and management. Lancet Neurol. 2009;8:192–204.
8. Kalra S, et al. Diagnosis and management of neuro-Behcet's disease: international consensus recommendations. J Neurol. 2014;261:1662–76. https://doi.org/10.1007/s00415-013-7209-3.
9. Serdaroglu P. Behcet's disease and the nervous system. J Neurol. 1998;245:197–205.
10. Siva A, et al. Behcet's disease: diagnostic and prognostic aspects of neurological involvement. J Neurol. 2001;248:95–103.
11. Hirohata S, Kikuchi H. Changes in biomarkers focused on differences in disease course or treatment in patients with neuro-Behcet's disease. Intern Med. 2012;51:3359–65.
12. Jongen PJ, Daelmans HE, Bruneel B, den Hartog MR. Humoral and cellular immunologic study of cerebrospinal fluid in a patient with Behcet encephalitis. Arch Neurol. 1992;49:1075–8.
13. Sharief MK, Hentges R, Thomas E. Significance of CSF immunoglobulins in monitoring neurologic disease activity in Behcets disease. Neurol. 1991;41:1398–401.
14. McLean BN, Miller D, Thompson EJ. Oligoclonal banding of IgG in CSF, blood-brain barrier function, and MRI findings in patients with sarcoidosis, systemic lupus erythematosus, and Behcet's disease involving the nervous system. J Neurol Neurosurg Psychiatry. 1995;58:548–54.
15. Saruhan-Direskeneli G, Yentur SP, Mutlu M, Shugaiv E, Yesilot N, Kurtuncu M, Akman-Demir G. Intrathecal oligoclonal IgG bands are infrequently found in neuro-Behcet's disease. Clin Exp Rheumatol. 2013;31:25–7.
16. Ohta M, Nishitani H, Matsubara F, Inaba G. Myelin basic protein in spinal fluid from patients with neuro-Behcet's disease. N Engl J Med. 1980;302:1093. https://doi.org/10.1056/NEJM198005083021922.
17. Kawai M, Hirohata S. Cerebrospinal fluid β2-microglobulin in neuro-Behçet's syndrome. J Neurol Sci. 2000;179:132–9. https://doi.org/10.1016/S0022-510X(00)00403-2.
18. Wang CR, Chuang CY, Chen CY. Anticardiolipin antibodies and interleukin-6 in cerebrospinal fluid and blood of Chinese patients with neuro-Behcet's syndrome. Clin Exp Rheumatol. 1992;10:599–602.

19. Hirohata S, Isshi K, Oguchi H, Ohse T, Haraoka H, Takeuchi A, Hashimoto T. Cerebrospinal fluid interleukin-6 in progressive neuro-Behcet's syndrome. Clin Immunol Immunopathol. 1997;82:12–7.
20. Akman-Demir G, et al. Interleukin-6 in neuro-Behçet's disease: association with disease subsets and long-term outcome. Cytokine. 2008;44:373–6. https://doi.org/10.1016/j.cyto.2008.10.007.
21. Saruhan-Direskeneli G, Yentur SP, Akman-Demir G, Isik N, Serdaroglu P. Cytokines and chemokines in neuro-Behcet's disease compared to multiple sclerosis and other neurological diseases. J Neuroimmunol. 2003;145:127–34.
22. Borhani Haghighi A, et al. CSF levels of cytokines in neuro-Behçets disease. Clin Neurol Neurosurg. 2009;111:507–10. https://doi.org/10.1016/j.clineuro.2009.02.001.
23. Gheita TA, Gamal SM, Shaker I, El Fishawy HS, El Sisi R, Shaker OG, Kenawy SA. Clinical significance of serum interleukin-23 and A/G gene (rs17375018) polymorphism in Behçets disease: relation to neuro-Behçet, uveitis and disease activity. Joint Bone Spine. 2015;82:213–5. https://doi.org/10.1016/j.jbspin.2014.10.008.
24. Niino M, Ogata A, Kikuchi S, Tashiro K, Nishihira J. Macrophage migration inhibitory factor in the cerebrospinal fluid of patients with conventional and optic-spinal forms of multiple sclerosis and neuro-Behçet's disease. J Neurol Sci. 2000;179:127–31. https://doi.org/10.1016/S0022-510X(00)00397-X.
25. Hamzaoui K, Hamzaoui A, Ghorbel I, Khanfir M, Houman H. Levels of IL-15 in serum and cerebrospinal fluid of patients with Behçet's disease. Scand J Immunol. 2006;64:655–60. https://doi.org/10.1111/j.1365-3083.2006.01844.x.
26. Hamzaoui K, Borhani-Haghighi A, Kaabachi W, Hamzaoui A. Increased interleukin 33 in patients with neuro-Behçet's disease: correlation with MCP-1 and IP-10 chemokines. Cell Mol Immunol. 2014;11:613. https://doi.org/10.1038/cmi.2014.31.
27. Belghith M, Bahrini K, Kchaou M, Maghrebi O, Belal S, Barbouche MR. Cerebrospinal fluid IL-10 as an early stage discriminative marker between multiple sclerosis and neuro-Behçet disease. Cytokine. 2018;108:160–7. https://doi.org/10.1016/j.cyto.2018.03.039.
28. Aldinucci A, et al. CSF/serum matrix metallopeptidase-9 ratio discriminates neuro Behçet from multiple sclerosis. Ann Clin Transl Neurol. 2018;5:493–8. https://doi.org/10.1002/acn3.538.
29. Hamzaoui K, Houman H, Hentati F, Hamzaoui A. BAFF is up-regulated in central nervous system of neuro-Behçet's disease. J Neuroimmunol. 2008;200:111–4. https://doi.org/10.1016/j.jneuroim.2008.06.006.
30. Sumita Y, Murakawa Y, Sugiura T, Wada Y, Nagai A, Yamaguchi S. Elevated BAFF levels in the cerebrospinal fluid of patients with neuro-Behçet's disease: BAFF is correlated with progressive dementia and psychosis. Scand J Immunol. 2012;75:633–40. https://doi.org/10.1111/j.1365-3083.2012.02694.x.
31. Hamzaoui K, Ayed K, Hamza MH, Hamzaoui A. VEGF and mRNA VEGF expression in CSF from Behçet's disease with neurological involvement. J Neuroimmunol. 2009;213:148–53. https://doi.org/10.1016/j.jneuroim.2009.05.005.
32. Gunaydin I, et al. The prevalence of Sjogren's syndrome in Behcet's syndrome. J Rheumatol. 1994;21:1662–4.
33. Yazici H. The place of Behcet's syndrome among the autoimmune diseases. Int Rev Immunol. 1997;14:1–10.
34. Direskeneli H. Autoimmunity vs autoinflammation in Behcet's disease: do we oversimplify a complex disorder? Rheumatology (Oxford). 2006;45:1461–5.
35. Direskeneli H, Saruhan-Direskeneli G. The role of heat shock proteins in Behcet's disease. Clin Exp Rheumatol. 2003;21:S44–8.
36. Lehner T. The role of heat shock protein, microbial and autoimmune agents in the aetiology of Behcet's disease. Int Rev Immunol. 1997;14:21–32.

37. Pervin K, et al. T cell epitope expression of mycobacterial and homologous human 65-kilodalton heat shock protein peptides in short term cell lines from patients with Behcet's disease. J Immunol. 1993;151:2273–82.
38. Nagafuchi H, et al. Excessive expression of Txk, a member of the Tec family of tyrosine kinases, contributes to excessive Th1 cytokine production by T lymphocytes in patients with Behcet/s disease. Clin Exp Immunol. 2005;139:363–70. https://doi.org/10.1111/j.1365-2249.2004.02688.x.
39. Direskeneli H, et al. Recognition of B-cell epitopes of the 65 kDa HSP in Behcet' s disease. Scand J Immunol. 1996;43:464–71.
40. Lule S, et al. Behçet disease serum is immunoreactive to neurofilament medium which share common epitopes to bacterial HSP-65, a putative trigger. J Autoimmun. 2017;84:87–96. https://doi.org/10.1016/j.jaut.2017.08.002.
41. Tasci B, Direskeneli H, Serdaroglu P, Akman-Demir G, Eraksoy M, Saruhan-Direskeneli G. Humoral immune response to mycobacterial heat shock protein (hsp)65 in the cerebrospinal fluid of neuro-Behcet patients. Clin Exp Immunol. 1998;113:100–4.
42. Celet B, et al. Anti-alpha B-crystallin immunoreactivity in inflammatory nervous system diseases. J Neurol. 2000;247:935–9.
43. Vural B, et al. Anti-neuronal and stress-induced-phosphoprotein 1 antibodies in neuro-Behçet's disease. J Neuroimmunol. 2011;239:91–7. https://doi.org/10.1016/j.jneuroim.2011.08.008.
44. Vural B, et al. Mitochondrial carrier homolog 1 (Mtch1) antibodies in neuro-Behçet's disease. J Neuroimmunol. 2013;263:139–44. https://doi.org/10.1016/j.jneuroim.2013.08.007.
45. Totsuka S, Midorikawa T. Some clinical and pathological problems in neuro-Behcet's syndrome. Folia Psychiatr Neurol Jpn. 1972;26:275–84.
46. Hirohata S. Histopathology of central nervous system lesions in Behcet's disease. J Neurol Sci. 2008;267:41–7. https://doi.org/10.1016/j.jns.2007.09.041.
47. Yamada H, Saito K, Hokari M, Toru S. Brain biopsy to aid diagnosis of neuro-Behçet's disease: case report and literature review. eNeurologicalSci. 2017;8:2–4. https://doi.org/10.1016/j.ensci.2017.06.001.
48. Frassanito MA, Dammacco R, Cafforio P, Dammacco F. Th1 polarization of the immune response in Behcet's disease: a putative pathogenetic role of interleukin-12. Arthritis Rheum. 1999;42:1967–74.
49. Imamura Y, et al. Involvement of Th1 cells and heat shock protein 60 in the pathogenesis of intestinal Behcet's disease. Clin Exp Immunol. 2005;139:371–8.
50. Lew W, Chang JY, Jung JY, Bang D. Increased expression of interleukin-23 p19 mRNA in erythema nodosum-like lesions of Behcet's disease. Br J Dermatol. 2008;158:505–11. https://doi.org/10.1111/j.1365-2133.2007.08403.x.
51. Nara K, Kurokawa MS, Chiba S, Yoshikawa H, Tsukikawa S, Matsuda T, Suzuki N. Involvement of innate immunity in the pathogenesis of intestinal Behcet's disease. Clin Exp Immunol. 2008;152:245–51. https://doi.org/10.1111/j.1365-2249.2008.03626.x.
52. Romagnani S. Human Th17 cells. Arthritis Res Ther. 2008;10:206. https://doi.org/10.1186/ar2392.
53. Hamzaoui K, Hamzaoui A, Guemira F, Bessioud M, Hamza M, Ayed K. Cytokine profile in Behcet's disease patients. Relationship with disease activity. Scand J Rheumatol. 2002;31:205–10.
54. Chi W, et al. Upregulated IL-23 and IL-17 in Behcet patients with active uveitis. Invest Ophthalmol Vis Sci. 2008;49:3058–64. https://doi.org/10.1167/iovs.07-1390.
55. Geri G, et al. Critical role of IL-21 in modulating TH17 and regulatory T cells in Behcet disease. J Allergy Clin Immunol. 2011;128:655–64. https://doi.org/10.1016/j.jaci.2011.05.029.
56. Nanke Y, Yago T, Kotake S. The role of Th17 cells in the pathogenesis of Behcet's disease. J Clin Med. 2017;6 https://doi.org/10.3390/jcm6070074.
57. Wang C, Tian Y, Ye Z, Kijlstra A, Zhou Y, Yang P. Decreased interleukin 27 expression is associated with active uveitis in Behcet's disease. Arthritis Res Ther. 2014;16:R117. https://doi.org/10.1186/ar4570.

58. Hamzaoui K, Borhani Haghighi A, Ghorbel IB, Houman H. RORC and Foxp3 axis in cerebrospinal fluid of patients with Neuro-Behçet's disease. J Neuroimmunol. 2011;233:249–53. https://doi.org/10.1016/j.jneuroim.2011.01.012.
59. Tulunay A, et al. Activation of the JAK/STAT pathway in Behcet's disease. Genes Immun. 2015;16:176. https://doi.org/10.1038/gene.2014.80.
60. Xavier JM, et al. Gene expression profiling and association studies implicate the neuregulin signaling pathway in Behcet's disease susceptibility. J Mol Med (Berl). 2013;91:1013–23. https://doi.org/10.1007/s00109-013-1022-4.
61. Puccetti A, et al. Gene expression profiling in Behcet's disease indicates an autoimmune component in the pathogenesis of the disease and opens new avenues for targeted therapy. J Immunol Res. 2018;2018:4246965. https://doi.org/10.1155/2018/4246965.
62. Bakir-Gungor B, Remmers EF, Meguro A, Mizuki N, Kastner DL, Gul A, Sezerman OU. Identification of possible pathogenic pathways in Behcet's disease using genome-wide association study data from two different populations. Eur J Hum Genet. 2015;23:678–87. https://doi.org/10.1038/ejhg.2014.158.
63. Yoshioka T, et al. Protein profiles of peripheral blood mononuclear cells as a candidate biomarker for Behcet's disease. Clin Exp Rheumatol. 2014;32:S9–19.
64. Melikoglu M, Kural-Seyahi E, Tascilar K, Yazici H. The unique features of vasculitis in Behcet's syndrome. Clin Rev Allergy Immunol. 2008;35:40–6. https://doi.org/10.1007/s12016-007-8064-8.
65. Kiraz S, Ertenli I, Ozturk MA, Haznedaroglu IC, Celik I, Calguneri M. Pathological haemostasis and "prothrombotic state" in Behcet's disease. Thromb Res. 2002;105:125–33.
66. Zierhut M, Mizuki N, Ohno S, Inoko H, Gul A, Onoe K, Isogai E. Immunology and functional genomics of Behcet's disease. Cell Mol Life Sci. 2003;60:1903–22.
67. Chamorro AJ, Marcos M, Hernandez-Garcia I, Calvo A, Mejia JC, Cervera R, Espinosa G. Association of allelic variants of factor V Leiden, prothrombin and methylenetetrahydrofolate reductase with thrombosis or ocular involvement in Behcet's disease: a systematic review and meta-analysis. Autoimmun Rev. 2013;12:607–16. https://doi.org/10.1016/j.autrev.2012.11.001.
68. Becatti M, et al. Neutrophil activation promotes fibrinogen oxidation and Thrombus formation in Behcet disease. Circulation. 2016;133:302–11. https://doi.org/10.1161/CIRCULATIONAHA.115.017738.
69. Yazici H, Seyahi E, Hatemi G, Yazici Y. Behcet syndrome: a contemporary view. Nat Rev Rheumatol. 2018;14:119. https://doi.org/10.1038/nrrheum.2018.3.
70. Buyon JP, Korchak HM, Rutherford LE, Ganguly M, Weissmann G. Female hormones reduce neutrophil responsiveness in vitro. Arthritis Rheum. 1984;27:623–30.
71. Yavuz S, Ozilhan G, Elbir Y, Tolunay A, Eksioglu-Demiralp E, Direskeneli H. Activation of neutrophils by testosterone in Behcet's disease. Clin Exp Rheumatol. 2007;25:S46–51.
72. Yavuz S, Akdeniz T, Hancer V, Bicakcigil M, Can M, Yanikkaya-Demirel G. Dual effects of testosterone in Behcet's disease: implications for a role in disease pathogenesis. Genes Immun. 2016;17:335–41. https://doi.org/10.1038/gene.2016.28.
73. Takeuchi M, et al. Dense genotyping of immune-related loci implicates host responses to microbial exposure in Behcet's disease susceptibility. Nat Genet. 2017;49:438–43. https://doi.org/10.1038/ng.3786.

Autoantibodies in Neuro-Behçet Disease

Yasemin Gürsoy-Özdemir

Behçet disease (BD) is a chronic, relapsing systemic disease that is characterized by inflammatory reaction involving mucocutaneous, ocular, joint, vessel, nervous, and gastrointestinal system. It is also known as "Silk Road disease," since its prevalence is very high in the Mediterranean and Asia than in Europe and the United States [1–4]. Turkey has the highest prevalence with 20–421/100,000 in adult population [1, 2].

Neuro-Behçet disease (NBD) is a subform of Behçet disease that mainly affects central nervous system. Both parenchymal and extra-parenchymal involvements were observed in patients. The most frequent and typical findings include focal parenchymal lesions, cerebral venous sinus thrombosis, arterial vasculitis, and meningo-encephalitic involvement especially affecting brain stem and extending from brain stem to basal structures like thalamus, hypothalamus, as well as internal capsule and basal ganglia [5]. In 5–15% of BD cases, neurological involvement occurs. Diagnosis is based on the international consensus recommendations [6], and there has been no laboratory diagnostic tool yet.

Although its etiology is unclear, Behçet disease is thought to be an autoinflammatory disease triggered by exogenous factors in genetically susceptible individuals [1, 4, 7–9]. Certain genotypic features like HLA-B51 allele are strongly associated with disease occurrence. Especially two alleles of HLA-B51, HLA-B5101 and HLA-B5108, have been found in close relation with BD [1, 3]. But genetic associations are not sufficient to explain the number of BD cases since HLA-B51 carriage rate is around 20% for genetic risk. Other than genetic susceptibility, immune mechanisms like innate and adaptive immunity triggered by environmental factors may have strong influential contribution to BD occurrence. It

Y. Gürsoy-Özdemir (✉)
Koç University School of Medicine, Department of Neurology and Graduate School of Health Sciences, İstanbul, Turkey
e-mail: ygursoy@ku.edu.tr

E. Tüzün, M. Kürtüncü (eds.), *Neuro-Behçet's Disease*,
https://doi.org/10.1007/978-3-030-55273-2_3

has been demonstrated that activation of the inflammasome and downstream inflammatory pathways play important roles in BD [3, 10].

Although BD is considered as an autoinflammatory disease, there are observations implying autoimmune etiology as well [1, 4, 7–9]. For example, autoinflammatory response to endothelial cells is thought to be one of the main reactions leading to vasculitis and other vascular involvement especially in neuro-Behçet disease patients [11]. But, other studies have also demonstrated the presence of antibodies against α-enolase on endothelial cells of Behçet disease patients. This enzyme is located on the surface of streptococci and hence may play a role in the induction of autoimmune disease caused by streptococci [12, 13].

BD has been categorized as autoinflammatory-autoimmune disorder, and several serologic markers have been studied up to now, and several studies have found an association with autoantibodies against the cellular structures that triggers inflammation and immune reactions like SIP-1, kinectin, intermediate filament, alpha tropomyosin, etc. [14–21].

In one study, Vural et al. detected antibodies against mitochondrial carrier homolog 1 (Mtch1) in Behçet disease patients' serum at a ratio of 47.2%. They have also demonstrated that disease severity and apoptotic cell death rates were positively correlated with antibody's association especially in NBD patients. They have argued that this finding is not a bystander side effect of auto-inflammation and possibly has some pathogenic function [22]. In line with this finding, antibodies against annexin V which is involved in apoptosis were significantly higher in BD patients compared to controls in another study [23]. In addition to its known role as an indicator of apoptosis, it could also be well thought of as a marker of disease activity. Authors proposed that anti-annexin antibodies could be considered as a risk for developing neurological manifestations in BD patients [20, 21, 23].

In another study, authors searched for anti-neuronal antibodies. They have incubated neuronal culture cells with BD and NBD patients' sera. They have detected the presence of autoantibodies to neuronal cell surface antigens in a proportion of NBD patients, greater than in BD controls (70% of NBD patients and 35% of BD patients' demonstrated positivity). They have additionally observed antibodies to stress-induced phosphoprotein-1 (STIP-1, 30% of NBD patients studied). Interestingly, higher titers of anti-STIP antibodies were detected at NBD patients' sera having parenchymal lesions [24]. Although the results are novel and interesting, authors recommended more detailed investigations, since STIP-1 antibodies have also been determined in ovarian cancer and rheumatoid arthritis patients [25, 26].

Distinct geographical representation of BD may be a clue for etiopathogenesis, and this distribution is proposed as an evidence for a genetic etiology [1–5]. On the other hand, this regional distribution may be an indication of the role of environmental factors, especially infections that had spread along this famous silk route. Hence, triggers of bacterial (mainly *Streptococcus sanguinis*) [27] or viral (mainly herpes simplex virus 1) [28] origin have long been considered in the pathogenesis of BD.

Bacterial etiology is also proposed as a possible mechanism for the BD pathophysiology [29]. Several studies have demonstrated that the ratio of *S. sanguinis* in the oral flora of BD is high compared to healthy controls [30, 31]. Interestingly, when T lymphocytes of BD patients are incubated with several bacteria, there was a general hyperactivity of BD patients' T lymphocytes against bacterial antigens and not only to S. sanguinis.

Cross immunopathology is supported by a mouse model showing symptoms similar to BD via inoculation of mice with BD serum. Especially, *Streptococcus sanguinis* and other bacteria found in oral mucosa of BD patients arc thought to be involved in this process via acting through the activation of heat shock and/or mechanical stress pathways [32]. Similarly, *Streptococcus sanguinis* was detected and grown in 58% of pustules in BD patients. On the other hand, only in 29% of pustules from acne vulgaris patients have contained *S. sanguinis* supporting the idea of bacterial involvement in BD pathogenesis [30].

Viral etiology is also proposed. First claim about viral etiology came from Hulusi Behçet. In 1937, he had observations of intracellular inclusion-like forms in smears from the hypopyon of the anterior chamber and from oral ulcers suggesting a viral etiology to BD [33]. A significant number of studies tried to correlate viral etiology with BD, especially herpes simplex virus-1 [34–37]. Later on with the development of novel techniques for viral detection, researchers re-examined for a viral etiology. Herpes simplex virus type 1 (HSV-1) DNA and RNA were detected in peripheral blood mononuclear cells of BD patients, especially patients with eye and joint involvement [35–37]. In addition to these observations, an animal model was developed by injecting mice with HSV [38]. These induced mucocutaneous symptoms include oral, genital, and skin ulcers; eye symptoms like uveitis, iridocyclitis, conjunctivitis, and hypopyon; gastrointestinal ulcers; arthritis; and skin crusting. However, unresponsiveness to antiviral agents and conflicting results with other viral etiologies diminish its direct role in disease pathophysiology [39].

Supporting bacterial and viral etiology, IgA antibodies to mycobacterial 65-kDa HSP, which is very similar to human 60-kDa HSP, were detected in high amounts in BD patients' sera when compared to controls [30, 40, 41].

Both bacterial and viral etiologies may be affecting the immune system through heat shock proteins (HSP). Bacterial HSP65 is actually very similar to human HSP60. This similarity may be causing cross-reactivity and induction of immune system, especially activation and proliferation of T lymphocytes in BD. Due to this resemblance, heat shock proteins are claimed as one of the main mechanisms for autoimmunity reactions [41, 42].

In line with the above-mentioned knowledge and data, it can be speculated that the association of infections with various autoimmune and inflammatory conditions points out to an altered immune response to infections in BD patients rather than active infection itself. Additionally, medical history of BD patients demonstrates chronic infection focuses like tonsillitis and dental caries when compared to healthy controls [30]. Hence, *Streptococcus sanguinis* is found as one of the main components of oral mucosa in BD patients. Additionally, serum antibody titers to *S. sanguinis* strains 113–20, 114–23, and 118–1 were also found in high amounts in BD patients when compared to healthy controls [30].

Although these studies point to viral or bacterial etiology, there is no direct pathway demonstrated especially for autoantigen studies. As we discussed, several autoantigens were detected in BD patients' sera, but none of the above-mentioned possible antibodies has been reported to have a high prevalence or share a common epitope with bacterial antigens like HSP65.

To find an association with HSP65, our group designed a delicate study for a novel autoantigen detection in BD sera. The presence of possible antibody was demonstrated by incubating mouse tissue sections with patient sera [10]. This method is used previously to detect possible autoantigens in other neurological diseases. One example is the detection of autoantibodies against aquaporin-4 channels in neuromyelitis optica.

Sera from 34 BD patients (14 with neurological involvement) were studied as well as those from patients with other neurological diseases like multiple sclerosis, neuromyelitis optica, and systemic lupus erythematosus. Mouse brain, eye, scrotal skin sections, and human hippocampal brain sections as well as primary neuron cultures were incubated with patient's sera and labelled indirectly with fluorescent-conjugated (FITC) goat anti-human IgG antibody. In all tissue sections, especially brain sections, fine filamentous immunolabelings were observed. These immunolabelings were mostly parenchymal, but some vascular involvements were also detected. Vascular involvement in all tissues was perivascular (labelling around vessels), and it was strongly associated with vascular neural innervations. These filamentous labeling by patient sera was highly co-localized with immunolabeling obtained by a commercially available anti-neurofilament-medium (NF-M) antibody that specifically labels NF-M protein (Fig. 1). Green labelled colon in A on the figure is NF-M staining of tissues, and red labelled colon is labelling of same tissues with patients' sera. It can be easily seen that both labelling resides on same areas. As the NF-M peptide is also expressed in peripheral axons within several tissues other than the CNS, immunoreactivity against filamentous neuronal processes was detected in retina and scrotal skin in majority of BD patients. This labelling may be a strong evidence for BD systemic involvements through NF-M. Additionally, labelling of perivascular nerves in all tissues examined may implicate that the antibodies to NF-M may be playing a role in vascular involvement as well.

All of these data were confirmed with incubation of BD patients' sera with human brain tissues and primary neuron cultures. To further analyze the structure of antibodies in BD patients' sera, we have incubated highly reactive sera with mouse brain tissue homogenates and western blot analyses were done. Positive bands that are not detected in control sera were analyzed by using matrix-assisted laser desorption/ionization-time of flight mass spectrometry (MALDI-TOF MS) and peptide mass fingerprinting (PMF). From this analysis, 37 target proteins are detected. Among these proteins, only NF-M, a neurofilament expressed in neuronal processes, was compatible with the filamentous neuropil staining observed with patient sera.

To make a correlation and cross reactivity against HSP65, we have further searched for epitope regions of *Streptococcus sanguinis* BD113–20 strain by Immune Epitope Database (IED).

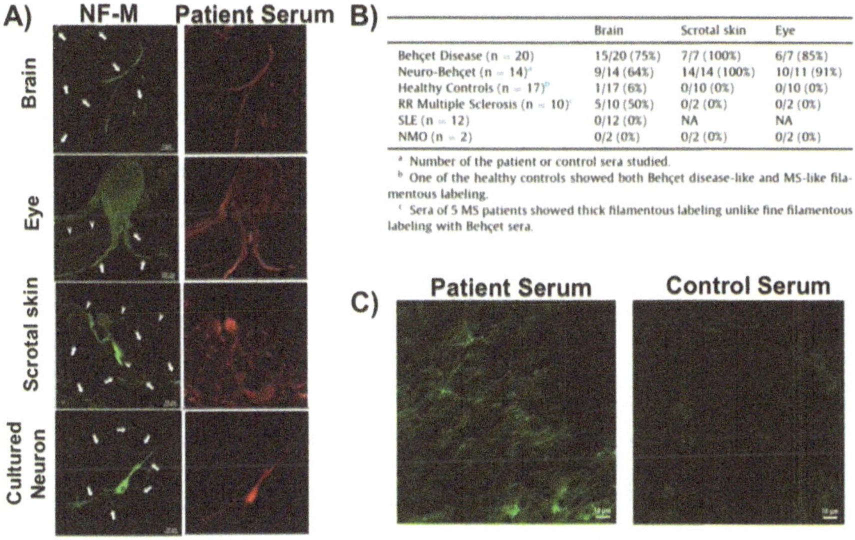

	Brain	Scrotal skin	Eye
Behçet Disease (n = 20)	15/20 (75%)	7/7 (100%)	6/7 (85%)
Neuro-Behçet (n = 14)[a]	9/14 (64%)	14/14 (100%)	10/11 (91%)
Healthy Controls (n = 17)[b]	1/17 (6%)	0/10 (0%)	0/10 (0%)
RR Multiple Sclerosis (n = 10)[c]	5/10 (50%)	0/2 (0%)	0/2 (0%)
SLE (n = 12)	0/12 (0%)	NA	NA
NMO (n = 2)	0/2 (0%)	0/2 (0%)	0/2 (0%)

[a] Number of the patient or control sera studied.
[b] One of the healthy controls showed both Behçet disease-like and MS-like filamentous labeling.
[c] Sera of 5 MS patients showed thick filamentous labeling unlike fine filamentous labeling with Behçet sera.

Fig. 1 Immunoreactivity of Behçet disease patient serum. (**a**) Serum incubation of mouse brain, eye, scrotal tissue, as well as neuronal culture cells produced same immunofluorescent staining pattern for NF-M. (**b**) Both BD and neuro-BD patients showed significantly high number of positive staining. (**c**) Note the filamentous staining pattern of mouse brain tissue with patient serum. (Reproduced with permission from Lule et al. [10])

Since there are lot of studies implying the association of HSPs with BD, we have found seven different epitopes of HSP65 as an antigen from IED. Interestingly, the 8th–18th amino acids (TGEWVNMIEEGIIDPVKV) of one of the antigen sequences demonstrated 45% identity with the 674–682nd amino acids of human NF-M (P07197) protein sequence as well as 55% identity with the 495–505th amino acids of M. tuberculosis HSP65 (P9WPE7) protein sequence. This homology may be the leading autoreactivity induced by bacterial HSPs and causing NF-M as target for immunological reactions.

Although these results point to an autoimmune etiology, it is evident that Behçet disease is a multifactorial disease. Further studies are granted to explain the pathophysiology and reactions against possible antibodies. If there will be strong indication for autoantibody as causative pathophysiology, then new and novel therapeutics can be developed.

References

1. Greco A, De Virgilio A, Ralli M, et al. Behçet disease: new insights into pathophysiology, clinical features and treatment options. Autoimmun Rev. 2018;17:567–75.
2. Onder M, Bulur I. Behçet disease: new aspects. Clin Dermatol. 2017;35:421–34.

3. De Chambrun MP, Weschler B, Geri G, et al. New insights into the pathogenesis of Behçet's disease. Autoimmun Rev. 2012;11:687–98.
4. Yazici H, Seyahi E, Hatemi G, Yazici Y. Behçet syndrome: a contemporary view. Nat Rev Rheumatol. 2018;108(14):107–19.
5. Casanova Peno I, De las Heras Revilla V, Parejo Carbonell B, et al. Neurobehçet disease: clinical and demographic characteristics. Eur J Neurol. 2012;19:1224–7.
6. Kalra S, Silman A, Akman-Demir G, et al. Diagnosis and management of Neuro-Behçet's disease: international consensus recommendations. J Neurol. 2014;261:1662–76.
7. Miller JJ, Venna N, Siva A. Neuro-Behçet disease and autoinflammatory disorders. Semin Neurol. 2014;34:437–43.
8. Takeuchi M, Mizuki N, Meguro A, et al. Dense genotyping of immune-related loci implicates host responses to microbial exposure in Behçet's disease susceptibility. Nat Genet. 2017;49(3):438–43.
9. Gül A. Pathogenesis of Behçet's disease: autoinflammatory features and beyond. Semin Immunopathol. 2015;37:413–8.
10. Lule S, Colpak AI, Balci-Peynircioglu B, et al. Behçet disease serum is immunoreactive to neurofilament medium which share common epitopes to bacterial HSP-65, a putative trigger. J Autoimmun. 2017;84:87–96. https://doi.org/10.1016/j.jaut.2017.08.002.
11. Christmann Souza R, Lage L, Goldesntein-Schainberg C, et al. Anti-endothelial cell antibodies and central nervous system involvement in Behçet's disease. Clinics. 2007;62(6):685–90. São Paulo. https://doi.org/10.1590/S1807-59322007000600005.
12. Hoon Lee K, Chung H-S, Sup Kim H, et al. Human α-enolase from endothelial cells as a target antigen of anti–endothelial cell antibody in Behçet's disease. Arthritis Rheum. 2003;48(7):2025–35.
13. Fontan PA, Pancholi V, Nociari MM, Fischetti VA. Antibodies to streptococcal surface enolase react with human_-enolase: implications in poststreptococcal sequelae. J Infect Dis. 2000;182:1712–21.
14. Akoglu T, Kozakoglu H, Akoglu E. Antibody to intermediate filaments of the cytoskeleton in patients with Behcet's disease. Clin Immunol Immunopathol. 1986;41(3):427–43.
15. Lu Y, Ye P, Chen SL, et al. Identification of kinectin as a novel Behcet's disease autoantigen. Arthritis Res Ther. 2005;7:R1133–9.
16. Mor F, Weinberger A, Cohen IR. Identification of alpha-tropomyosin as a target self-antigen in Behcet's syndrome. Eur J Immunol. 2002;32:356–65.
17. Vural B, Demirkan A, Ugurel E, et al. Seroreactivity against PTEN-induced putative kinase 1 (PINK1) in Turkish patients with Behcet's disease. Clin Exp Rheumatol. 2009;27:S67–72.
18. Briani C, Doria A, Marcolongo R, et al. Increased titres of IgM anti-heparan sulfate antibody in Behcet's disease. Clin Exp Rheumatol. 2006;24:S104–7.
19. Delunardo F, Conti F, Margutti P, et al. Identification and characterization of the carboxy-terminal region of Sip-1, a novel autoantigen in Behcet's disease. Arthritis Res Ther. 2006;8:R71.
20. Aslan H, Pay S, Gok F, et al. Anti Annexin V autoantibody in thrombophilic Behcet's disease. Rheumatol Int. 2004;24:77–9.
21. Chen P, Yan H, Tian Y, Xun Y, Shi L, Bao R, et al. Annexin A2 as a target endothelial cell membrane autoantigen in Behcet's disease. Sci Rep. 2015;5:8162.
22. Vural B, Şehitoğlu E, ÇavuŞ F, et al. Mitochondrial carrier homolog 1 (Mtch1) antibodies in neuro-BehÇet's disease. J Neuroimmunol. 2013;263:139–44.
23. Gheita TA, Samir H, Hussein H. Anti-annexin V antibodies in neuro-Behçmet patients: clinical significance and relation to disease activity. Int J Rheum Dis. 2012;15:e124–6.
24. Vural B, Uğurel E, Tüzün E. Anti-neuronal and stress-induced-phosphoprotein 1 antibodies in neuro-Behçet's disease. J Neuroimmunol. 2011;239:91–7.
25. Goëb V, Thomas-L'Otellier M, Daveau ea. Candidate autoantigens identified by mass spectrometry in early rheumatoid arthritis are chaperones and citrullinated glycolytic enzymes. Arthritis Res Ther. 2009;11:R38.

26. Kim S, Cho H, Nam EJ, Kim SW, Kim YT, Park YW, Kim BW, Kim JH. Autoantibodies against stress-induced phosphoprotein-1 as a novel biomarker candidate for ovarian cancer. Genes Chromosomes Cancer. 2010;49:585–95.
27. Lehner T, Lavery E, Smith R, van der Zee R, Mizushima Y, Shinnick T. Association between the 65-kilodalton heat shock protein, Streptococcus sanguis, and the corresponding antibodies in Behcet's syndrome. Infect Immun. 1991;59:1434–41.
28. Nomura Y, Kitteringham N, Shiba K, Goseki M, Kimura A, Mineshita S. Use of the highly sensitive PCR method to detect the Herpes simplex virus type 1 genome and its expression in samples from Behcet disease patients. J Med Dent Sci. 1998;45:51–8.
29. Marta M, Santos E, Coutinho E, et al. The role of infections in Behçet disease and neuro-Behçet syndrome. Autoimmun Rev. 2015;14:609–15.
30. Hatemi G, Yazıcı H. Behçet's syndrome and micro-organisms. Best Pract Res Clin Rheumatol. 2011;25:389–406.
31. Yoshikawa K, Kotake S, Sasamoto Y, Ohno S, Matsuda H. Close association of Streptococcus sanguis and Behçet's disease. Nippon Ganka Gakkai Zasshi. 1991;95(12):1261–7.
32. Isogai E, Isogai H, Kotake S, Ohno S, Kimura K, Oguma K. Role of streptococcus sanguis and traumatic factors in Behçet's disease. J Appl Res. 2003;3:64–75.
33. Behçet H. Über rezidivierende Aphtösedurch Ein Virus verursachte Geschwüre am Mund, am Auge und an der Genitalien. Derm Wschr. 1937;105(36):1152–63.
34. Sezer F. The isolation of a virus as the cause of Behçet's disease. Am J Ophthalmol. 1953;36:301–6.
35. Eglin RP, Lehner T, Subak-Sharpe JH. Detection of RNA complementary to herpes-simplex virus in mononuclear cells from patients with Behcet's syndrome and recurrent oral ulcers. Lancet. 1982;2(8312):1356–61.
36. Studd M, McCance DJ, Lehner T. Detection of HSV-1 DNA in patients with Behcet's syndrome and in patients with recurrent oral ulcers by the polymerase chain reaction. J Med Microbiol. 1991;34(1):39–43.
37. Lee S, Bang D, Cho YH, Lee ES, Sohn S. Polymerase chain reaction reveals herpes simplex virus DNA in saliva of patients with Behcet's disease. Arch Dermatol Res. 1996;288:179–83.
38. Sohn S, Bang D, Lee ES, Kwon HJ, Lee SI, Lee S. Experimental studies on the antiviral agent famciclovir in Behçet's disease symptoms in ICR mice. Br J Dermatol. 2001;145:799–804.
39. Davies UM, Palmer RG, Denman AM. Treatment with acyclovir does not affect orogenital ulcers in Behcet's syndrome: a randomized double-blind trial. Br J Rheumatol. 1988;27(4):300–2.
40. Lehner T, Lavery E, Smith R, van der Zee R, Mizushima Y, Shinnick T. Association between the 65-kilodalton heat shock protein, Streptococcus sanguis, and the corresponding antibodies in Behcet's syndrome. Infect Immunol. 1991;59(4):1434–41.
41. Direskeneli H, Hasan A, Shinnick T, Mizushima R, van der Zee R, Fortune F, et al. Recognition of B-cell epitopes of the 65 kDa HSP in Behcet's disease. Scand J Immunol. 1996;43(4):464–71.
42. Kaneko F, Oyama N, Yanagihori H, Isogai E, Yokota K, Oguma K. The role of streptococcal hypersensitivity in the pathogenesis of Behcet's disease. Eur J Dermatol. 2008;18:489–98.

Animal Models of Behçet's Disease

Hande Yüceer and Erdem Tüzün

Introduction

Behçet's disease (BD) is an episodic, inflammatory, vascular disorder distinctively characterized by recurrent uveitis, aphthae, genital ulcerations, and skin and eye lesions. It may affect multiple systems in the body including vascular, musculoskeletal, gastrointestinal, ocular, and nervous systems (neuro-Behçet disease) [1–3]. Inflammation, environmental factors, viral and bacterial infections, and genetics have been proposed as major etiological factors leading to BD onset.

Animal models have been developed in order to gain better understanding of BD etiopathogenesis and progression as well as to test new therapeutic approaches which cannot be applied on human subjects. Although the complex nature of the disease makes the ideal experimental design harder to achieve, recent animal studies have shown similar disease pathology, symptoms, and treatment response to BD [4]. These animal models might be classified as infection-induced, genetic, immunological, and environmental models depending on the causative etiology. This part will focus on these experimental animal designs of BD in the light of recent literature.

H. Yüceer (✉) · E. Tüzün
Department of Neuroscience, Aziz Sancar Institute of Experimental Medicine, Istanbul University, Istanbul, Turkey

E. Tüzün, M. Kürtüncü (eds.), *Neuro-Behçet's Disease*,
https://doi.org/10.1007/978-3-030-55273-2_4

Infection-Induced Models

Herpes Simplex Virus-Induced Model

Since the first description of BD by Hulusi Behçet in the early twentieth century, viral infection has been hypothesized as one of the putative etiological aspects of the disease. By providing evidence of research throughout the years, diverse kinds of viruses in varied BD cases have been detected. In those studies, human herpesvirus types 6 and 7, herpes simplex virus (HSV), human immunodeficiency virus, hepatitis virus, cytomegalovirus, varicella zoster virus, parvovirus B19, and Epstein-Barr virus have been isolated from bodily fluids or tissues of different BD patients [5]. For instance, the virus from hypopyon fluid, which was first doubted but could not be named by Hulusi Behçet himself, has been primarily isolated from BD patients and transfused to the central nervous system of the mice, the guinea pigs, and the rabbits by Sezer in 1953. As a result of that inoculation, symptoms like encephalitis and thrombophlebitis (mostly in the mice), roughened coat, immobilization or hyperactivity, convulsions, tremor, and eye lesions (especially in the rabbits) have been observed [6, 7]. Interestingly, the histopathologic signs of neurodegeneration related to BD on mice have also shown as glial nodules [6]. In 1969, another team from Istanbul has induced meningoencephalitis and NBD-like symptoms, caused by inoculating the cerebrospinal fluid (CSF) of an NBD patient into mice [8].

Many other researchers have tried to detect the genetic material of herpes simplex virus generally in the peripheral blood mononuclear cells (PBMC) of the BD patients and focused on the relationship between HSV and the occurrence of different BD symptomatology [9–13]. On the grounds of those studies, among all viral infectious models, herpes simplex virus (HSV) model has become the most reliable method to induce BD in experimental animals [14]. Initially, it has been observed by Sohn et al. that HSV-immunized mice had manifestations resembling BD such as eye lesions and genital and oral ulcerations [15]. Neurological involvement has also been detected in subsequently developed HSV-induced models [16].

Furthermore, as noted by Baharav et al. [4], ICR (Institute of Cancer Research) strain mice are frequently preferred in viral infectious animal designs of BD. In addition, the clinical manifestations of BD could be induced in 40–50% of C57BL/6, B10.RIII, and B10.BR, whereas only in 2% of BALB/c and C3H/He strains [4, 17].

Even though the mortality rate has been high, HSV-induced model has been validated in terms of reliability and constancy of mimicking BD-like symptoms and histopathological and immunological features of the disease after inoculation as well as generating therapeutic response similar to BD patients [5, 14]. However, the available evidence suggests that anti-herpesvirus treatment has remained ineffective for BD. Therefore, HSV infection has been considered not as a direct cause of the disease but as a secondary factor triggering inflammatory dysregulation due to immune response to a viral antigen [4]. The ample support for the immunopathological premise has been provided by disparate therapeutic studies.

Streptococcal Infection Model

In addition to viral infection, bacterial infection has also been suggested to be one of the etiological factors behind BD. Especially the role of oral hygiene in BD symptoms has been debated. *Streptococcus* genus bacteria have been found significantly higher in oral bacterial flora of BD patients. In bacterial infectious animal models, the different kinds of *Streptococcus* genus bacteria (*Streptococcus faecalis*, *Streptococcus pyogenes*, *Streptococcus sanguinis*, *Streptococcus salivarius*) have been isolated from BD patients and injected into the experimental animals. As a consequence of that model, septic shock, short-term uveitis, and acute multisystemic inflammatory reactions have been observed [18]. The most prominent indication provided by that model has been the eye involvement.

Moreover, lipoteichoic acid (LTA), which is a streptococcal antigen found in the oral flora of BD patients, has also been suggested as an animal model for BD depending on LTA antibodies' impact on pro-inflammatory IL-8 cytokine level in rats [19]. Nevertheless, bacterial infectious models have never been a prevailing technique in BD [4].

Genetic Models

The current literature on BD pays particular attention to its genetic etiology and the determining role of human leukocyte antigen (HLA)-B51 in BD susceptibility. With this in mind, preliminary work on transgenic animal model of BD has been undertaken by Takeno et al. in 1995. Once the HLA-B51 gene has been inserted into C3H/He mice, the rate of peripheral blood neutrocyte activation and superoxide production has gone up. Surprisingly, none of the BD symptoms could have been developed in that model [20]. Although the results of transgenic model have not supported the dominant predisposing function of HLA-B51 in BD, that discrepancy could be attributed to insufficiency of HLA-B51 to induce BD-like manifestations or the possibly counteractive lineage of the mice [4, 17, 20]. Hence, it has also been conceivably hypothesized by Mizuki et al. that MHC class I polypeptide-related sequence A (MICA) could have been a new predisposing gene associated with BD more than HLA-B51 [21].

Immunological Models

Heat Shock Protein Model

Heat shock proteins (HSP) are molecular chaperons regulating immune functions under stress condition. As efficacious activators of innate immune system, HSP are responsible for pro-inflammatory cytokine production, maintaining Th1/Th2 balance and inducing CD4+/CD8+ cell responses [22]. In BD, human HSP60 and bacterial HSP65 have been found to play a role in increased B and T cell responses [7, 22]. When HSP-derived peptide is orally or subcutaneously administered to rats, clinical symptoms of BD with the inclusion of uveitis, recurrent oral aphthae, thrombosis, and neurological manifestations have been observed [4, 22]. CD8+ cells have appeared to intercept, while CD4+ cells have interceded uveitis in HSP model. Moreover, the CD24+ and CD25+ T cells prevent harmful immune response and autoimmunity. The transfusion of CD4+ CD25+ T cells has modified the levels of interleukin (IL)-10, IL-6, and IL-17 and ameliorated the BD-mimicking symptoms in mice [23]. On these grounds, HSP model could be evaluated as an autoimmune model for BD [4, 24].

Anti-neuronal Antibody Model

Recently, anti-neuronal antibodies have been thought of as a key factor especially in NBD cases [25]. Based on that premise, Sprague-Dawley rats have been immunized with anti-neuronal immunoglobulin G (IgG) isolated from NBD patients' sera. The most striking result of the experiment has been diminished locomotor activity in the open field test. The reason why motor symptoms occurred has been debated to be stemming from the neurodegenerative function of neuropil antibodies expressed by corticospinal tract fibers [26]. That model has been the first to show NBD-like manifestations by anti-neuronal antibody inducement in an experimental animal design.

Tropomyosin Model

Tropomyosin, a protein regulating actin filaments and contraction of muscle cells, has been shown to be another agent for autoimmune BD model. Several studies have revealed that when inoculated with alpha-tropomyosin (TPM), Lewis rats have begun to suffer from uveitis, arthritis, and other commmon indications of BD [27]. It seems probable that these symptoms are due to the auto-antigen role of alpha-TPM accompanied by Th1 cytokine profile and increased levels of tumor necrosis factor-alpha (TNF-alpha) [4, 27].

Retinal S-Antigen Model

One of the most acknowledged proteins for retinal vasculitis and autoimmune uveitis of BD has been known as retinal S-Antigen (S-Ag) located in the rod, photoreceptor area of the eye. Immunized rats with S-Ag have been reported to develop uveitis through increased CD8+ activation. Because of the fact that S-Ag induction has caused only inflammatory eye disease, this model might be classified as an organ-specific autoimmune BD model [4, 28].

Environmental Models

The Pitman-Moore miniature swine model has been the historically first described animal model of BD. The basis of this model has been to expose experimental animals to environmental pollutants (inorganic copper, organophosphate, and organic chloride) for an extended period. As a result of this exposure, BD-like manifestations such as oral aphthae, genital ulcerations, and folliculitis have been induced [14]. Although Pitman-Moore miniature pig model has created highly similar BD-mimicking symptoms, it has never been a preferable practice due to the difficulty of generating the model and the longer time needed (4–10 months) to begin seeing the symptoms. Notwithstanding the limitations of that model, it has been an enlightening method to become aware of the environmental substances' crucial role in immunological regulation [4].

Therapeutic Studies of BD Animal Models

In the past two decades, the therapeutics including colchicine, azathioprine, thalidomide, cyclosporin, aciclovir, gemcitabine, vitamin D3, and many others have been tested mostly on HSV-induced BD animal models [5]. For example, while famciclovir administration in HSV-induced mice model has improved oral, genital, and eye lesions and lowered mortality rate with accompanied decreased level of IL-2, thalidomide treatment improved mucocutaneous symptoms by upregulating FasR, perforin, and MIP-1a and downregulating TNF-alpha [16, 29]. BD-like symptoms have also been reduced when TNF-alpha expression has been decreased by small interfering RNAs (siRNAs) [30]. Besides, pentoxifylline or colchicine has caused the fall of microRNA-21 (miRNA-21) level associated with HSV-induced inflammation and healed the clinical signs of BD in mice [31]. On the other hand, polycytidylic acid (poly I:C), as an immunostimulant, has been reported to improve BD-like symptoms in HSV-infected mice by upregulating IL-15R-alpha+ and memory T cells while downregulating IL-17A [32].

In another study, BD-resembling clinical manifestations have been ameliorated by the downregulation of Toll-like receptor 2 (TLR2) and Toll-like receptor 4 (TLR4) expression [33]. Further, galectin-9 (Gal-9), the ligand of T cell immunoglobulin and mucin domain-containing molecule 3 (TIM-3), has mended inflammation in HSV-infectious BD model [34]. Additionally, TIM-1 vector and TIM-4 siRNA treatments have upregulated regulatory T cells as well as derogated symptom severity in BD mice [35]. All in all, a growing body of literature has shown that BD animal models would help us investigate not only the effectiveness of therapeutic treatments, and the improvement in clinical symptomatology but also discover new immunopathological mechanisms of the disease.

Conclusion

Collectively, these infection-induced, genetic, immunological, and environmental models provide BD etiopathogenesis with precious insights. However, among all experimental animal designs, HSV-induced model has been the most reliable and widely used one in terms of producing highly similar multisystemic BD characteristics [14]. The conclusion can be drawn from all BD animal designs that viral infection model has been more determinative to stimulate BD-like manifestations, especially in mice, compared to genetic susceptibility models. Since, what we know about genetic BD animal models has been greatly based upon HLA-B51; MICA model of BD might be developed as an alternative promising animal model of the disease. Also, these models should be tested on divergent mice strains including B10.BR, B10.RIII, and C57BL/6 in the future [17]. Likewise, the potentiality of TPM model as a convenient autoimmune BD animal design due to its significance in both innate and adaptive immune systems needs further evaluation [4].

Above all, the generalizability of all these models into NBD has been the major limitation. Apart from early viral and bacterial infectious models, a few exceptional HSP-derived BD models, and recent anti-neuronal antibody model of NBD, neurological involvement and NBD-like symptoms are quite arduous to induce in experimental animals [6, 8, 16, 18, 22, 26]. Further NBD animal studies are required so as to clarify the exact etiology of neurological manifestations in BD and to advance novel efficient therapeutic approaches for both BD and NBD in the future.

References

1. Hirohata S, Kikuchi H. Behçet's disease. Arthritis Res Ther. 2003;5:139–46. https://doi.org/10.1186/ar757.
2. International Study Group for Behçet's Disease. Criteria for diagnosis of Behcet's disease. Lancet. 1990;335(8697):1078–80. https://doi.org/10.1016/0140-6736(90)92643-V.
3. Saip S, Akman-Demir G, Siva A. Neuro-Behcet syndrome. Handb Clin Neurol. 2014;121:1703–23. https://doi.org/10.1016/b978-0-7020-4088-7.00110-3.

4. Baharav E, Weinberger A, Mor F, Krause I. Animal models of Behçet's disease. In: Yazıcı Y, Yazıcı H, editors. Behçet's Syndrome: Springer; 2010. p. 277–84. https://doi.org/10.1111/j.1476-5381.2011.01386.x.
5. Kim DY, Cho S, Choi MJ, Sohn S, Lee ES, Bang D. Immunopathogenic role of herpes simplex virus in Behcet's disease. Genet Res Int. 2013;2013:638273. https://doi.org/10.1155/2013/638273.
6. Sezer FN. The isolation of a virus as the cause of Behçet's disease. Am J Ophthalmol. 1953;36(3):301–15. https://doi.org/10.1016/0002-9394(53)91372-5.
7. Yildirim O. Animal models in Behçet's disease. Pathol Res Int. 2012;2012:273701. https://doi.org/10.1155/2012/273701.
8. Noyan B, Gürsoy G, Aktin E. Inoculation of cerebrospinal fluid of a neuro-Behçet-patient into mice. Acta Neuropathol. 1969;12(2):195–9. https://doi.org/10.1007/BF00692507.
9. Eglin R, Lehner T, Subak-Sharpe J. Detection of Rna complementary to herpes-simplex virus in mononuclear cells from patients with Behçet's syndrome and recurrent oral ulcers. Lancet. 1982;320(8312):1356–61.
10. Evans AD, Pallis CA, Spillane JD. Involvement of the nervous system in Behcet's syndrome. Report of three cases and isolation of virus. Lancet. 1957:349–53. Retrieved from https://www.cabdirect.org/cabdirect/abstract/19582700772.
11. Lee S, Bang D, Cho YH, Lee ES, Sohn S. Polymerase chain reaction reveals herpes simplex virus DNA in saliva of patients with Behçet's disease. Arch Dermatol Res. 1996;288(4):179–83.
12. Mortada A, Imam IZ. Virus aetiology of Behçet's syndrome. Br J Ophthalmol. 1964;48:250–9. Retrieved from http://www.ncbi.nlm.nih.gov/pubmed/14193679.
13. Studd M, Mccance DJ, Lehner T. Detection of HSV-1 DNA in patients with BehCet's syndrome and in patients with recurrent oral ulcers by the polymerase chain reaction. J Med Microbiol. 1991;34:39–43.
14. Sohn S, Lee ES, Bang D. Learning from HSV-infected mice as a model of Behcet's disease. Clin Exp Rheumatol. 2012;30(10):96–103. https://doi.org/10.1111/j.1346-8138.2012.01621.x.
15. Sohn S, Lee ES, Bang D, Lee S. Behçet's disease-like symptoms induced by the Herpes simplex virus in ICR mice. Eur J Dermatol. 1998;8(1):21–3. Retrieved from http://www.ncbi.nlm.nih.gov/pubmed/9649665.
16. Sohn S, Bang D, Lee E-S, Kwon HJ, Lee SI, Lee S. Experimental studies on the antiviral agent famciclovir in Behcet's disease symptoms in ICR mice. Br J Dermatol. 2001;145(5):799–804. https://doi.org/10.1046/j.1365-2133.2001.04498.x.
17. Sohn S, Lee ES, Lee S. The correlation of MHC haplotype and development of Behçet's disease-like symptoms induced by herpes simplex virus in several inbred mouse strains. J Dermatol Sci. 2001;26(3):173–81. https://doi.org/10.1016/S0923-1811(01)00086-X.
18. Kaneko F, Oyama N, Nishibu A. Streptococcal infection in the pathogenesis of BehÇet's disease and clinical effects of minocycline on the disease symptoms. Yonsei Med J. 1997;38(6):444. https://doi.org/10.3349/ymj.1997.38.6.444.
19. Cuchacovich M, Merino G, Yamamoto JH, Villarroel F, Saavedra T, Jofre S, Gatica H, Velasquez V, Pizzo SV, Gonzalez-Gronow M. Behçet's disease patients present high levels of deglycosylated anti-lipoteichoic acid IgG and high IL-8 production after lipoteichoic acid stimulation. Clin Exp Rheumatol. 2005;23(4–38):27–34.
20. Takeno M, Kariyone A, Yamashita N, Takiguchi M, Mizushima Y, Kaneoka H, Sakane T. Excessive function of peripheral blood neutrophils from patients with Behcet's disease and from HLA-B51 transgenic mice. Arthritis & Rheumatism. 1995;38(3):426–433.
21. Mizuki N, Ota M, Kimura M, Ohno S, Ando H, Katsuyama Y, Bahram S. Triplet repeat polymorphism in the transmembrane region of the MICA gene: a strong association of six GCT repetitions with Behcet disease. Proceedings of the National Academy of Sciences. 1997;94(4):1298–1303.
22. Direskeneli H, Saruhan-Direskeneli G. The role of heat shock proteins in Behçet's disease. Clin Exp Rheumatol. 2003;21(4–30):44–8.

23. Shim J, Lee E-S, Park S, Bang D, Sohn S. CD4(+) CD25(+) regulatory T cells ameliorate Behcet's disease-like symptoms in a mouse model. Cytotherapy. 2011;13(7):835–47. https://doi.org/10.3109/14653249.2011.571245.
24. Hu W, Hasan A, Wilson A, Stanford MR, Li-Yang Y, Todryk S, Whiston R, Shinnick T, Mizushima Y, van der Zee R, Lehner T. Experimental mucosal induction of uveitis with the 60-kDa heat shock protein-derived peptide 336–351. Eur J Immunol. 1998;28:2444–55.
25. Giriş M, Bireller S, Küçükali Cİ, Hanağasi H, Değirmencioğlu S, Tüzün E. Impact of neuro-Behçet disease immunoglobulin G on neuronal apoptosis. Noro Psikiyatr Ars. 2017;54(1):67–71. https://doi.org/10.5152/npa.2016.19421.
26. Erdağ E, Şahin C, Küçükali Cİ, Bireller S, Küçükerden M, Kürtüncü M, Türkoğlu R, Cakmakoglu B, Tüzün E, Arıcıoğlu F. Effects of in vivo and in vitro administration of neuro-Behcet's disease IgG. Neurol Sci. 2017;38(5):833–43. https://doi.org/10.1007/s10072-017-2856-2.
27. Baharav E, Mor F, Halpern M, Quintana F, Weinberger A. Tropomyosin-induced arthritis in rats. Clin Exp Rheumatol. 2007;25(4–45):86–92.
28. Ermakova NA, Alekberova ZS, Prokaeva TB. Autoimmunity to S-antigen and retinal vasculitis in patients with Behçet's disease. In: Adamantiades-Behçet's disease. Boston: Kluwer Academic Publishers; 2003. p. 279–81. https://doi.org/10.1007/0-306-48382-3_56.
29. Lee ES, Kim YA, Kwon HJ, Bang D, Lee S, Sohn S. Thalidomide upregulates macrophage inflammatory protein-1α in a herpes simplex virus-induced Behçet's disease-like animal model. Arch Dermatol Res. 2004;296(4):175–81. https://doi.org/10.1007/s00403-004-0498-8.
30. Choi B, Hwang Y, Kwon HJ, Lee ES, Park KS, Bang D, Lee S, Sohn S. Tumor necrosis factor alpha small interfering RNA decreases herpes simplex virus-induced inflammation in a mouse model. J Dermatol Sci. 2008;52(2):87–97. https://doi.org/10.1016/j.jdermsci.2008.05.001.
31. Choi B, Kim HA, Suh CH, Byun HO, Jung JY, Sohn S. The relevance of miRNA-21 in HSV-induced inflammation in a mouse model. Int J Mol Sci. 2015;16(4):7413–27. https://doi.org/10.3390/ijms16047413.
32. Choi J, Lee ES, Choi B, Sohn S. Therapeutic potency of Poly I: C in HSV-induced inflammation through up-regulation of IL-15 receptor alpha. Immunobiology. 2013;218(9):1119–30. https://doi.org/10.1016/j.imbio.2013.03.005.
33. Choi B, Lee ES, Sohn S. Vitamin D3 ameliorates herpes simplex virus-induced Behçet's disease-like inflammation in a mouse model through down-regulation of Toll-like receptors. Clin Exp Rheumatol. 2011;29(67):13–9.
34. Shim JA, Park S, Lee ES, Niki T, Hirashima M, Sohn S. Galectin-9 ameliorates herpes simplex virus-induced inflammation through apoptosis. Immunobiology. 2012;217(6):657–66. https://doi.org/10.1016/j.imbio.2011.11.002.
35. Shim JA, Lee ES, Choi B, Sohn S. The role of T cell immunoglobulin mucin domains 1 and 4 in a herpes simplex virus-induced Behçet's disease mouse model. Mediat Inflamm. 2013;2013:903948. https://doi.org/10.1155/2013/903948.

Pathology

Gökçen Ünverengil

Behçet's disease (BD) is a chronic relapsing inflammatory disease that affects many organs. Its major pathologic feature is vasculitis. Neurological involvement in Behçet's disease is relatively rare. Berlin reported the first autopsy case of BD involving the central nervous system (CNS) [1]. Almost all known entities about neuro-Behçet's disease (NBD) are based on reports of autopsy cases.

There are two main pathologic subtypes of NBD. The most common form is *parenchymal NBD*, and the second form is *non-parenchymal NBD*, which is also called as neuro-vasculo-BD. Occurrence of these two forms in the same individual is so rare [2, 3].

Parenchymal NBD

Parenchymal involvement is believed to be due to inflammatory small vessel disease and is mainly seen at the brain stem, diencephalon, and basal ganglia. Spinal cord, hemispheric, cerebellar, or meningeal involvement may also be seen [4–8]. It may present as acute disease or may have a chronic progressive form.

Macroscopical Findings.

The gross pathological change in chronic NBD is the atrophy of the brain stem with formation of cystic lesions. Especially the cerebral peduncles and the basis pontis are usually atrophic. Atrophy of the spinal cord and the discolorations of the pyramidal tracts and fasciculus gracilis have also been reported. Fibrous thickening of the meninges may be seen [1, 5, 8–11]. There can be cerebral atrophy with mild or prominent enlargement of ventricles [12], but brain stem atrophy in the absence of cortical atrophy is rather specific for NBD [13]. At the cut surface, there may be small softening areas and necrotic foci [1, 4–6, 11, 14].

Intracranial hemorrhage [15] and pseudotumoral presentation [16–18] may be seen rarely.

G. Ünverengil (✉)
Istanbul University, Istanbul Faculty of Medicine, Department of Pathology, Istanbul, Turkey
e-mail: gokcen.gundogdu@istanbul.edu.tr

E. Tüzün, M. Kürtüncü (eds.), *Neuro-Behçet's Disease*,
https://doi.org/10.1007/978-3-030-55273-2_5

Histological Findings

Histopathologically, BD has no pathognomonic features. It has been considered that CNS lesions in BD are caused by vasculitis with venous predominance [6, 14]. However, since the lesions do not usually have features of fibrinoid necrosis or infiltration of inflammatory cells in the vessel walls, it is indicated to be a perivasculitis rather than a true vasculitis [8, 9, 19]. However, fibrinoid necrosis was also reported to be observed in a few studies [6, 20].

Morphological findings in the acute phase lesions are characterized by infiltration of lymphocytes and monocytes around small vessels with or without necrosis. Polymorphonuclear leukocytes and rarely eosinophils may be seen in the parenchyma [5, 8, 21–24]. Immunohistochemical studies have shown that the mononuclear cell population consists of predominantly T lymphocytes and histiocytes and there were a few B lymphocytes [21, 22]. In chronic progressive lesions, similar histopathological changes were noted, although the degree of mononuclear cell infiltration was modest. There were also scattered foci of neurons undergoing apoptosis with formation of a few binucleated neurons both in acute and chronic NBD. It is suggested that soluble factors produced by infiltrating cells, including IL-6, might play a role in the induction of apoptosis of neurons in NBD. In late stages, demyelination and gliosis are the predominant histopathologic features and small inflammatory attacks thought to take place during the remission phase [6, 21].

Perivascular sudanophilic foam-cell infiltration has been described as one of the characteristic features of NBD [14]. Hirohata showed that these foam-cells were immunohistochemically CD68 positive, so they may be activated macrophages, which phagocyte the damaged white matters (myelin) [21].

The histological changes may also be seen in the optic nerves [7, 23].

Non-parenchymal NBD

Non-parenchymal involvement of NBD is mostly seen as a venous pathology [23] which includes dural sinus thrombosis, arterial occlusion, and arterial aneurysms [25]. The superior sagittal sinus is the most common site of thrombosis, followed by the transverse sinuses, deep cerebral veins, and cavernous sinuses. Arterial involvement is a rare cause of neurovasculo-BD which can present as stenosis, aneurysm formation, or dissection of the cerebral arteries. These pathologies may cause intracerebral and/or subarachnoid hemorrhage [23].

Peripheral Nervous System Involvement

Peripheral nervous system involvement is relatively rare in BD [3, 23]. Histopathologically, lymphocytic infiltration within the nerve bundles and nerve infarction, axonal degeneration and regeneration, or Wallerian degeneration may be seen [23, 26, 27] (Fig. 1).

a

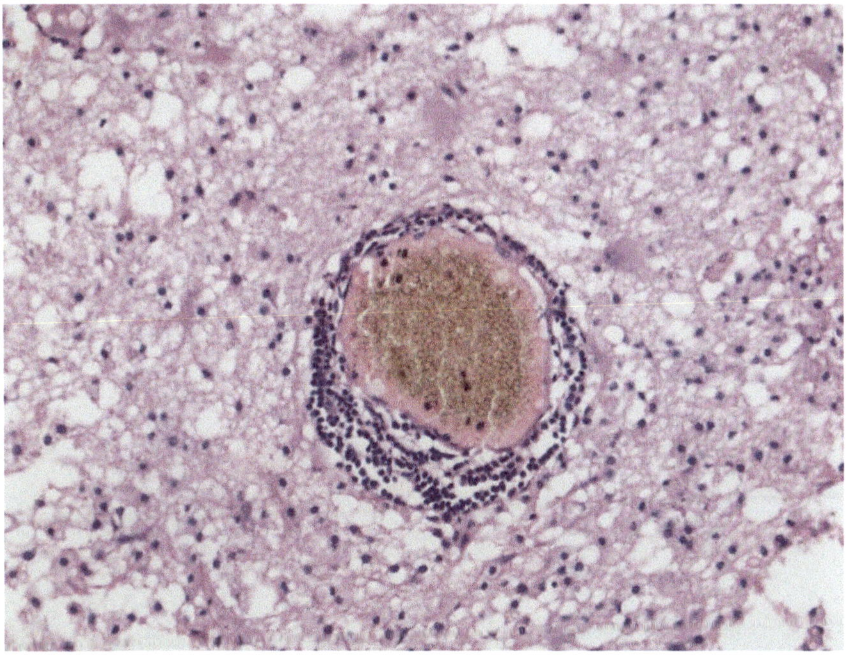

b

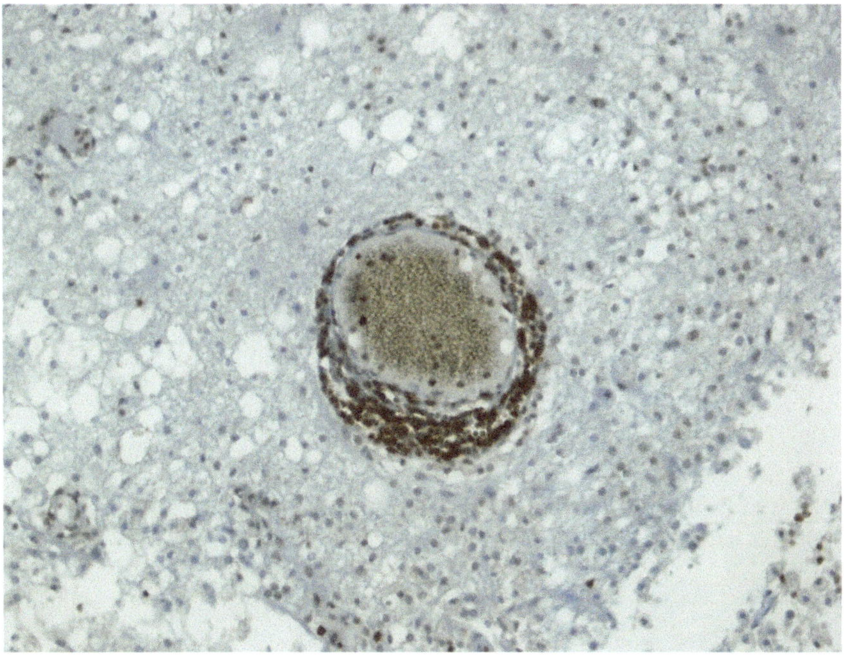

Fig. 1 (**a**) Mononuclear infiltration around a medium-sized vein in the brain parenchyma. There is no fibrinoid necrosis, thrombosis, or infiltration of the vessel wall. There are foam cells around the vessel wall and in the brain parenchyma (HE×200). (**b**, **c**) The infiltration consists of mostly CD45RO-positive T lymphocytes (**b**), and there are very few CD20-positive B lymphocytes (**c**) (×200). (**d**) Foam cells are immunohistochemically positive for CD68 (×200)

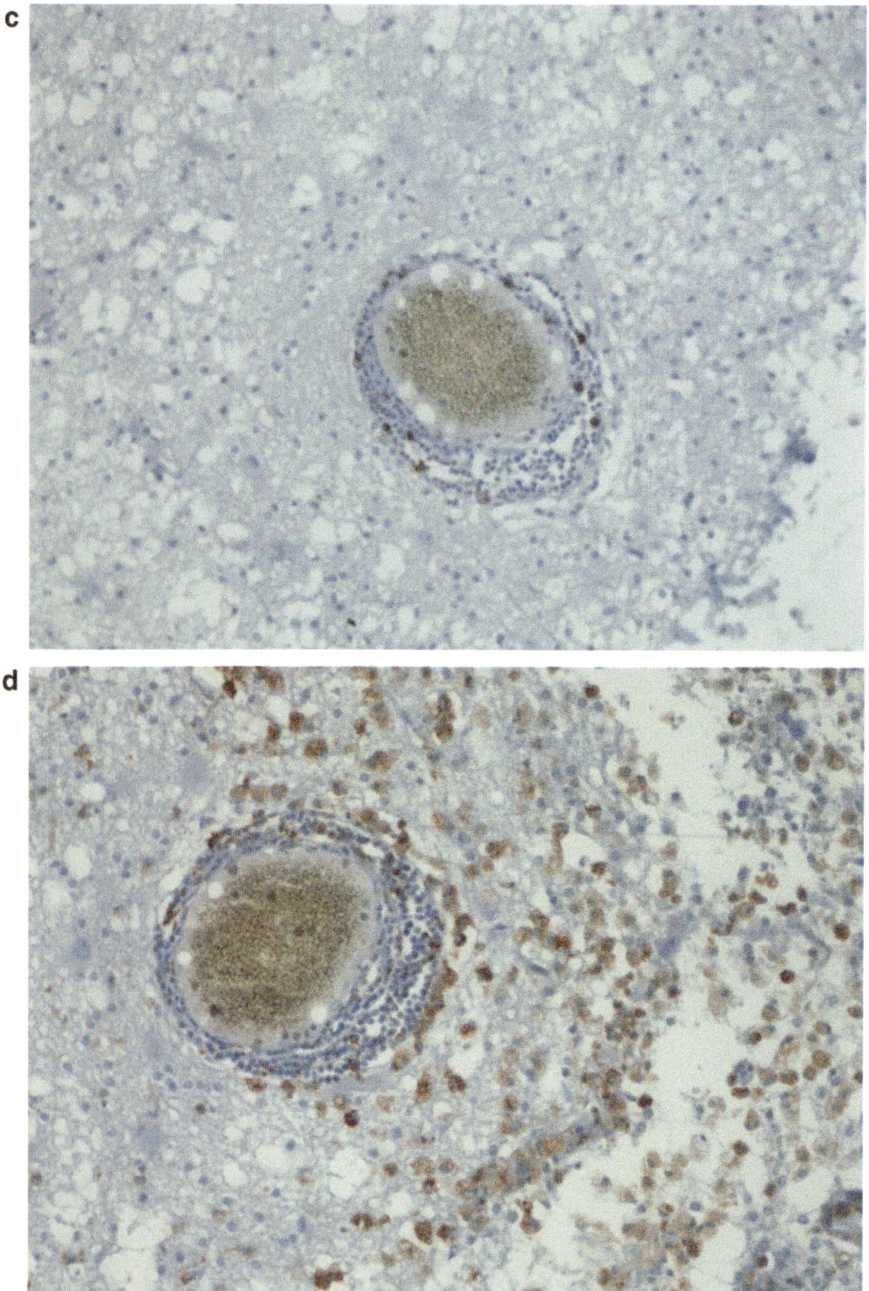

Fig. 1 (continued)

References

1. Berlin C. Behçet's syndrome with involvement of central nervous system. Report of a case, with necropsy, of lesions of the mount, genitalia and eyes; review of the literature. Arch Derm Syph (Chic). 1944;49:227–33.
2. Akman-Demir G, Serdaroglu P, Tasci B. Clinical patterns of neurological involvement in Behçet's disease: evaluation of 200 patients. The Neuro-Behçet's Study Group. Brain. 1999;122:2171–82.
3. Saip S, Akman-Demir G, Siva A. Neuro-Behçet syndrome. Handb Clin Neurol. 2014;121:1703–23.
4. Silfverskiold BP. Recurrent uveitis (Behçet syndrome) and encephalomyelomeningitis. Acta Psychiatr Neurol Scand. 1951;26:443–53.
5. Totsuka S, Hattori T, Yazaki M, et al. Clinicopathologic studies on neuro-Behcet's disease. Folia Psychiatr Neurol Jpn. 1985;39:155–66.
6. Kawakita H, Nishimura M, Satoh Y, et al. Neurological aspects of Behcet's disease. A case report and clinico-pathological review of the literature in Japan. J Neurol Sci. 1967;5:417–39.
7. Lakhanpal S, Tani K, Lie JT, Katoh K, Ishigatsubo Y, Ohokubo T. Pathologic features of Behçet's syndrome: a review of Japanese autopsy registry data. Hum Pathol. 1985;16(8):790–5.
8. Arai Y, Kohno S, Takahashi Y, et al. Autopsy case of neuro-Behcet's disease with multifocal neutrophilic perivascular inflammation. Neuropathology. 2006;26:579–85.
9. Riera-Mestre A, Martınez-Yelamos S, Martinez-Yelamos A, et al. I. Clinicopathologic features and outcomes of neuro-Behcet disease in Spain: a study of 20 patients. Eur J Intern Med. 2010;21(6):536–41.
10. Kaneko H, Nakajima H, Okamura A, et al. Histopathology of Behçet disease. Pathol Int. 1976;26(6):765–79.
11. Sugihara H, Muto Y, Tsuchiyama H. Neuro-Behcet's syndrome: report of two autopsy cases. Acta Pathol Jpn. 1969;19:95–101.
12. Miyakawa T, Murayama E, Deshimaru M, et al. Neuro-Behcet's disease showing severe atrophy of the cerebrum. Acta Neuropathol. 1976;34:95–103.
13. Çoban O, Bahar S, Akman-Demir G, Tasçi B, Yurdakul S, Yazici H, Serdaroglu P. Masked assessment of MRI findings: is it possible to differentiate neuro-Behçet's disease from other central nervous system diseases? Neuroradiol. 1999;41:255–60.
14. McMenemey W, Lawrence BJ. Encephalomyelopathy in Behçet's disease; report of necropsy findings in two cases. Lancet. 1957;273:353–8.
15. Kikuchi S, Niino M, Shinpo K, et al. Intracranial hemorrhage in neuro-Behcet's syndrome. Intern Med. 2002;41:692–5.
16. Noel N, Hutie M, Wechsler B, et al. Pseudotumoural presentation of neuro-Behçet's disease: case series and review of literature. Rheumatology. 2012;51:1216–25.
17. Yamada H, Saito K, Hokari M, et al. Brain biopsy to aid diagnosis of neuro-Behçet's disease: case report and literature review. eNeurologicalSci. 8:2–4.
18. Appenzeller S, de Castro R, LdS Q, et al. Brain tumor-like lesion in Behçet disease. Rheumatol Int. 2006;26(6):577–80.
19. Hadfield MG, Aydin F, Lippman HR, et al. Neuro-Behçet's disease. Clin Neuropathol. 1997;16(2):55–60.
20. Scardamaglia L, Desmond PM, Gonzales MF, et al. Behçet's disease with cerebral vasculitis. Intern Med J. 2001;31(9):560–1.
21. Hirohata S. Histopathology of central nervous system lesions in Behcet's disease. J Neurol Sci. 2008;267:41–7.
22. Hirohata S, Arai H, Matsumoto T. Immunohistological studies in neuro-Behçet's disease. Adv Exp Med Biol. 2003;528:385–7.
23. Borhani Haghighi A, Pourmand R, Nikseresht AR. Neuro-Behcet disease. A review. Neurologist. 2005;11(2):80–9.

24. Kalra S, Silman A, Akman-Demir G, et al. Diagnosis and management of neuro-Behçet's disease: inter- national consensus recommendations. J Neurol. 2014;261(9):1662–76.
25. Borhani Haghighi A, Ashjazadeh N. Acute disseminated encephalomyelitis-like manifestations in a patient with neuro-Behçet disease. Neurologist. 2009;15(5):282–4.
26. Namer IJ, Karabudak R, Zilelı T, et al. Peripheral nervous system involvement in Behçet's disease. Eur Neurol. 1987;26(4):235–40.
27. Walker LJ, Swallow MW, Mirakhur M. Behçet's disease presenting with mononeuritis multiplex. Ulster Med J. 1990;59(2):206–10.

Behçet's Disease: Clinical Features

Fatma Alibaz-Oner and Haner Direskeneli

Introduction

Behçet's disease (BD) is a chronic, multisystemic, inflammatory disease characterized by recurrent attacks of mucocutaneous, ocular, musculoskeletal, vascular, central nervous system (CNS), and gastrointestinal (GI) manifestations. The disease was first described by Hulusi Behçet, a Turkish dermatologist, as a triple complex with oral, genital ulcers and hypopyon uveitis in 1937 [1]. BD has a disease course with remission and relapses; complete remission is observed in at least 60% of the patients at 20 years [2] (Table 1).

The course of BD is more severe in males and in patients having a young disease onset (<25 years old) [3, 4]. The disease most frequently starts in the second and third decades of life. The disease onset can be in pediatric ages, but onset after 50 years is very rare [5–7]. Some of the manifestations of BD also show regional differences. Gastrointestinal findings are common in Japanese and Korean patients, but rather infrequent in patients from Turkey and the Middle East. In contrast, vascular disease is rare in East Asia. Pathergy test, the nonspecific hypersensitivity of the skin to a needle prick, is less commonly positive in North European and North American Caucasian patients [8].

F. Alibaz-Oner · H. Direskeneli (✉)
Marmara University, School of Medicine Hospital, Department of Internal Medicine, Division of Rheumatology, İstanbul, Turkey
e-mail: haner@marmara.edu.tr

E. Tüzün, M. Kürtüncü (eds.), *Neuro-Behçet's Disease*,
https://doi.org/10.1007/978-3-030-55273-2_6

Table 1 International criteria for classification of Behçet's disease

Recurrent oral ulceration *Minor aphthous, major aphthous, or herpetiform ulceration observed by a physician or reported reliably by a patient* *Recurrent at least three times in one 12-month period*
Plus 2 of:
Recurrent genital ulceration Recurrent genital aphthous ulceration or scarring, especially males, observed by a *physician or reported reliably by a patient*
Eye lesions *Anterior uveitis* *Posterior uveitis* *Cells in vitreous on slit lamp examination or retinal vasculitis observed by a qualified physician*
Skin lesions *Erythema nodosum-like lesions observed by a physician or reported reliably by a patient* *Pseudo-folliculitis* *Papulopustular lesions or acneiform nodules consistent with BD*
Positive pathergy test *An erythematous papule,>2 mm, at the prick site after the application of a sterile needle, 20–22 gauge, which obliquely penetrated avascular skin to a depth of 5 m; read by a physician at 48 h*

Note: Findings are applicable if no other clinical explanation is present

Clinical Features

Mucocutaneous Involvement

Recurrent Aphthous Ulcers

Recurrent aphthous (oral) ulcers are seen in 95–97% of patients and are usually the first disease manifestation of BD, preceding the diagnosis by an average of 6–7 years. They often present as painful, erythematous, circular, and slightly raised areas evolving into oval or round ulcers within 48 h (Fig. 1). Oral ulcers are frequently observed on the mucous membranes of the lips, gingiva, cheeks, and tongue, perhaps more posteriorly than ordinary aphthae. They usually heal in about 10–15 days without scarring [9–12]. In a long-term routine follow-up, oral aphthous ulcers were observed to be the main cause of ongoing clinical activity [13].

Genital Ulcers

Genital ulcers (GU) are another major manifestations of BD. They are also the most specific (95%) mucocutaneous sign of BD [14]. The frequency of genital ulcers changes between 50 and 85%. They are usually located in the scrotum in males and in the major and minor labiae in females (Fig. 2). GU usually begin as papules or

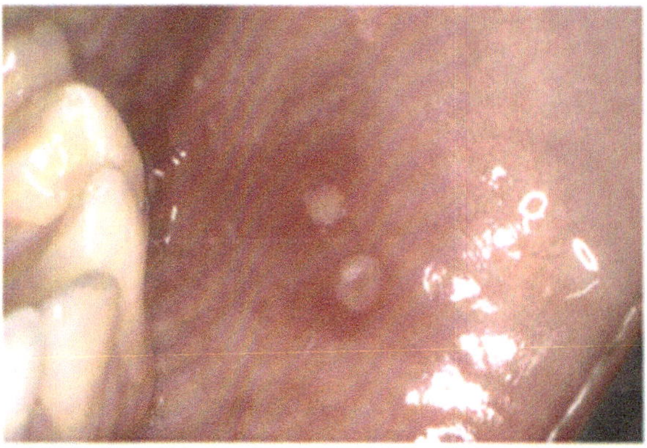

Fig. 1 Oral ulcers in Behcet's disease. (Courtesy of Prof. Dr. Tulin Ergun)

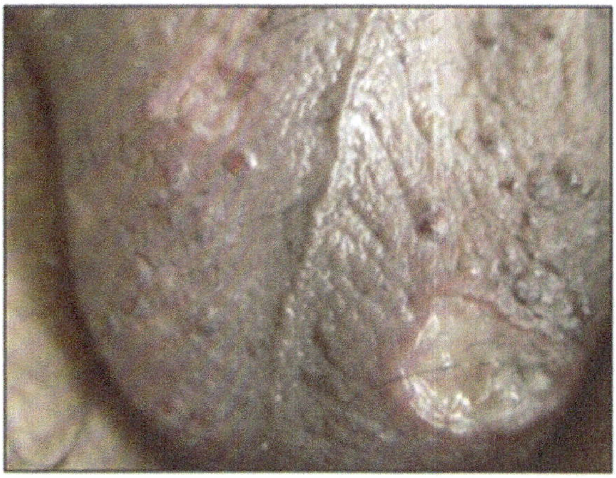

Fig. 2 Genital ulcer in Behcet's disease. (Courtesy of Prof. Dr. Tulin Ergun)

pustules that ulcerate after a short time [14–17]. GU usually heal in 10–30 days if they are not secondarily infected and leave scars in about 60% of the patients [18].

Cutaneous Lesions

Other types of cutaneous lesions in BD can be grouped into three categories [19]:

1. Erythema nodosum (EN)-like lesions and superficial thrombophlebitis
2. Papulopustular and acneiform lesions
3. Other lesions such as skin ulcers and Sweet syndrome

Erythema nodosum-like lesions are red, tender, erythematous non-ulcerating nodules which are frequently located on the legs and generally resolve with pigmentation [14–16]. EN-like lesions are observed in approximately 50% of BD patients and are more frequently present in females [20]. Superficial thrombophlebitis (ST) is the most common type of venous involvement [21], and it can be difficult to

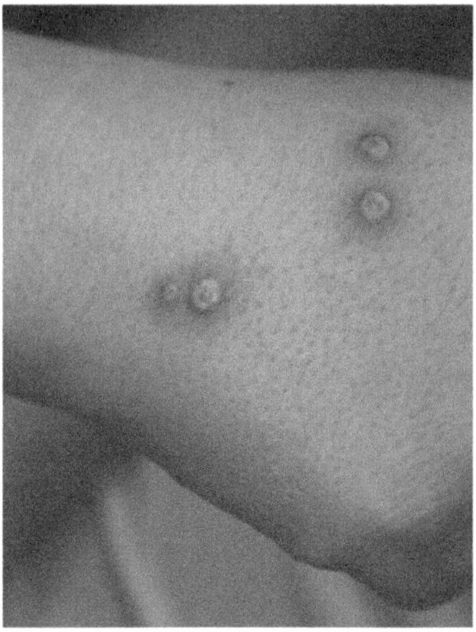

Fig. 3 Papulo-pustuler lesions in Behcet's disease. (Courtesy of Prof. Dr. Tulin Ergun)

differentiate ST and EN in some cases. It presents as palpable, painful subcutaneous nodules which are string-like hardenings with reddening of the overlying skin [22].

Papulopustular lesions are frequently indistinguishable from ordinary acne. They are seen at usual acne sites such as face, upper chest, and back and additionally on the legs and arms [23] (Fig. 3). Other cutaneous lesions such as skin ulcers and Sweet syndrome can also be observed in BD.

Pathergy Reaction

Pathergy test is a nonspecific hyperreactivity in response to minor trauma in the skin and is quite specific for BD. It is usually used as a diagnostic test and is performed by inserting a 20–22-gauge needle into the dermis of the forearm of the patient (Fig. 4a). The presence of a papule or pustule at 48 h is considered positive (Fig. 4b) [24].

Eye Involvement

Eye involvement is one of the main causes of morbidity in BD. It is observed in up to 50% of patients. Generally, a chronic, relapsing, bilateral uveitis is present. Ocular inflammation is commonly panuveitis and retinitis. However, some patients can present with an isolated anterior uveitis. Retinal lesions consist of exudates, hemorrhages, papilledema, and macular disease. Post-inflammatory changes such as synechia and retinal scars are important determinants of prognosis in eye

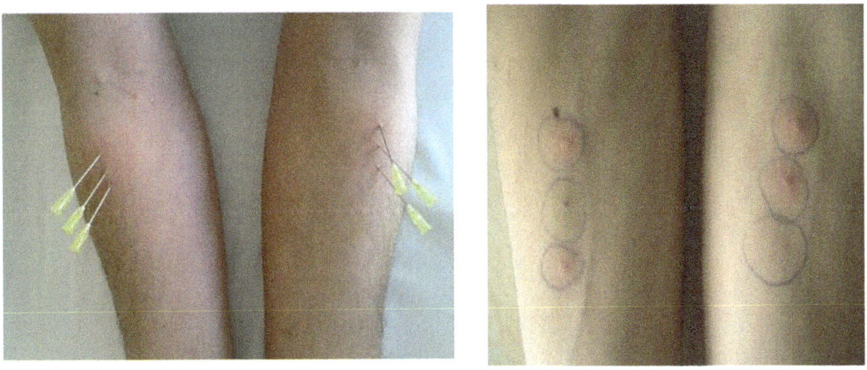

Fig. 4 (**a**, **b**) Pathergy test. (Courtesy of Prof. Dr. Tulin Ergun)

involvement [8]. In the 20-year Cerrahpaşa Outcome Survey, eye involvement was bilateral in 80% of the males and 64% of females at the first visit. At the end of 20 years of follow-up, 87% of males and 71% of females had bilateral eye disease [2].

Musculoskeletal Disease

Arthritis or arthralgia is seen in about 50% of patients with BD. It is usually manifested as a non-deforming, non-erosive peripheral oligoarthritis in decreasing order of the knees, ankles, hands, and wrists. It usually resolves in a couple of days to weeks [25]. Sacroiliitis is not a prominent part of the clinical picture, but the coexistence of acne, arthritis, and enthesopathies suggests that at least a subgroup of patients have reactive arthritis-like features [20].

Vascular Involvement

Vasculitis is a main pathological finding in BD. Vessels of all sizes can be involved, both in the arterial and venous systems [26, 27]. Major vessel involvement consists of arterial occlusion, arterial aneurysms, and major vein occlusions (Fig. 5). Vascular involvement is seen in up to 40% of the patients with BD, especially in young males, and is one of the major causes of mortality and morbidity. Venous involvement (80%) is reported to be more common than arterial disease [10]. Lower extremity vein thrombosis is the most frequent form of vascular involvement [2] (Fig. 6). Although venous thrombosis is seen primarily in the lower extremities, it may affect many different sites including inferior and superior vena cava, pulmonary artery, suprahepatic vessels, and cardiac cavities. Up to 17% of the mortality in

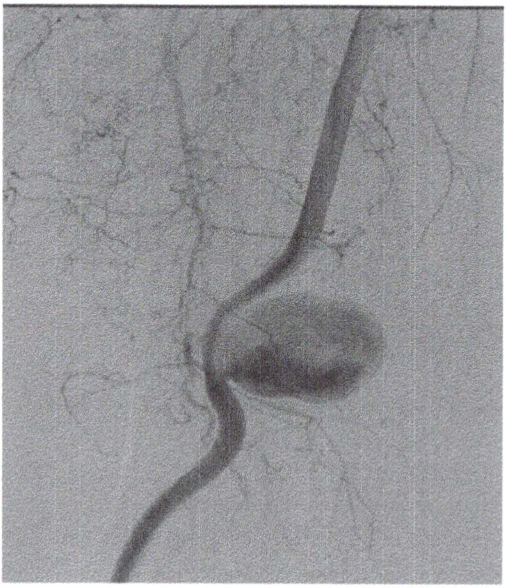

Fig. 5 A pseudoaneurysm in right superficial femoral artery in BD

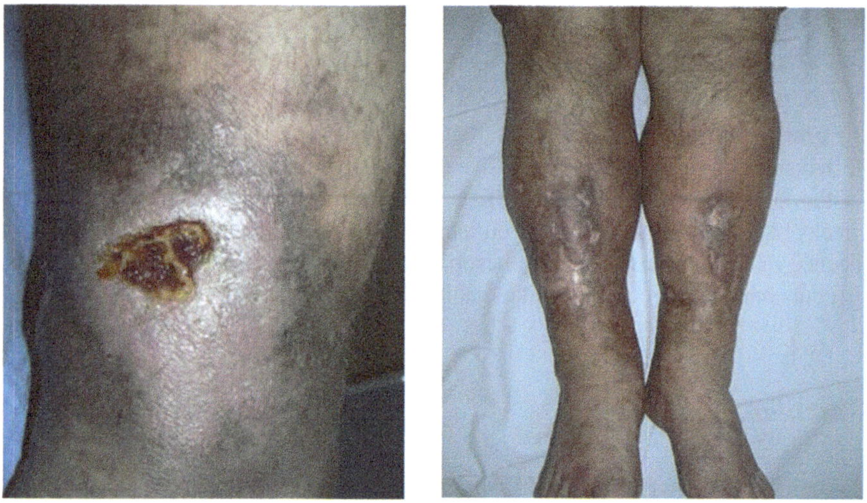

Fig. 6 Lower extremity venous insufficiency and ulcers. (Courtesy of Prof. Dr. Tulin Ergun)

BD is reported to be associated with venous involvement such as pulmonary embolism or Budd-Chiari syndrome [28]. There have been sporadic reports of valvular lesions, myocarditis, coronary vasculitis, ventricular aneurysms, and intracavitary thrombus formation, but overall cardiac involvement is uncommon in BD.

Gastrointestinal Involvement

Gastrointestinal involvement is seen in one-third of BD patients in Japan and Korea but is rare in Mediterranean countries (<5%). It is characterized by mucosal ulcerations primarily in the terminal ileum and the cecum. The main symptoms are vomiting, abdominal pain, and diarrhea. A mass is often palpable in the abdomen during exacerbations, and ileocecal perforations may rarely occur. It is sometimes difficult to distinguish the findings of gastrointestinal BD from those of Crohn's disease [29]. Younger age at diagnosis is associated with a more severe disease course and a poorer prognosis [14].

Other Clinical Findings

Renal involvement such as glomerulonephritis is rarely reported as sporadic cases in BD. AA-type amyloidosis can occasionally be seen [15] As a cause of testicular pain, epididymitis can be observed in male patients [16]. A recent study from Turkey also reported increased incidence of varicocele in Behçet's disease [17].

References

1. Behçet H. Uber rezidiverende, aphthose, durch ein Virus verursachte Gescgwure am Mund, um Auge und an den Genitalen. Dermatol Wochenschri. 1937;105:1152–7.
2. Kural-Seyahi E, Fresko I, Seyahi N, Ozyazgan Y, Mat C, Hamuryudan V, et al. The long-term mortality and morbidity of Behcet syndrome: a 2-decade outcome survey of 387 patients followed at a dedicated center. Medicine (Baltimore). 2003;82:60–76.
3. O'Neill TW, Silman AJ, Rigby AS, et al. Sex and regional differences in clinical manifestations of Behçet's disease. Br J Rheumatol. 1993;32(Suppl 1):85.
4. Yazici H, Tüzün Y, Pazarli H, Yurdakul S, Ozyazgan Y, Ozdoğan H, et al. Influence of age of onset and patient's sex on the prevalence and severity of manifestations of Behçet's syndrome. Ann Rheum Dis. 1984;43(6):783–9.
5. Barnes CG. History and diagnosis. In: Yazıcı Y, Yazıcı H, editors. Behçet's disease. New York/Dordrecht/Heidelberg/London: Springer; 2010. p. 7–34.
6. Shafaie N, Shahram F, Davatchi F, et al. Behçet's disease in children. In: Wechsler B, Godeau P, editors. Behçet's disease, International congress series 1037. Amsterdam: Excerpta Medica; 1993. p. 381–3.
7. Zouboulis CC, Kötter I, Djawari D, Kirch W, Kohl PK, Ochsendorf FR, et al. Epidemiological features of Adamantiades-Behçet's disease in Germany and in Europe. Yonsei Med J. 1997;38(6):411–22.
8. Fresko F, Stübiger N, Tascilar K. Behçet's Syndrome, relapsing polychondritis and eye involvement in rheumatik disease (Modül 25). Eular On-line Course on Rheumatic Diseases (2007–2012).
9. Idil A, Gürler A, Boyvat A, Caliskan D, Ozdemir O, Isik A, et al. The prevalence of Behçet's disease above the age of 10 years. The results of a pilot study conducted at the Park Primary Health Care Center in Ankara, Turkey. Ophthalmic Epidemiol. 2002;9(5):325–31.

10. Alpsoy E, Donmez L, Onder M, Gunasti S, Usta A, Karincaoglu Y, et al. Clinical features and natural course of Behçet's disease in 661 cases: a multicentre study. Br J Dermatol. 2007;157(5):901–6. Epub 2007 Aug 17.
11. Saylan T, Mat C, Fresko I, Melikoğlu M. Behçet's disease in the Middle East. Clin Dermatol. 1999;17(2):209–23; discussion 105–6.
12. Bang D, Hur W, Lee ES, Lee S. Prognosis and clinical relevance of recurrent oral ulceration in Behçet's disease. J Dermatol. 1995;22(12):926–9.
13. Alibaz-Oner F, Mumcu G, Kubilay Z, Ozen G, Celik G, Karadeniz A, et al. Unmet need in Behcet's disease: most patients in routine follow-up continue to have oral ulcers. Clin Rheumatol. 2014;33(12):1–7. https://doi.org/10.1007/s10067-014-2585-3.
14. Park JJ, Kim WH, Cheon JH. Outcome predictors for intestinal Behçet's disease. Yonsei Med J. 2013;54(5):1084–90. https://doi.org/10.3349/ymj.2013.54.5.1084.
15. Melikoğlu M, Altiparmak MR, Fresko I, Tunç R, Yurdakul S, Hamuryudan V, et al. A reappraisal of amyloidosis in Behçet's syndrome. Rheumatology (Oxford). 2001;40(2):212–5.
16. Cetinel B, Akpinar H, Tüfek I, Uygun N, Solok V, Yazici H. Bladder involvement in Behçet's syndrome. J Urol. 1999;161(1):52–6.
17. Yilmaz O, Yilmaz S, Kisacik B, Aydogdu M, Bozkurt Y, Erdem H, et al. Varicocele and epididymitis in Behcet disease. J Ultrasound Med. 2011;30(7):909–13.
18. Mat MC, Goksugur N, Engin B, Yurdakul S, Yazici H. The frequency of scarring after genital ulcers in Behçet's syndrome: a prospective study. Int J Dermatol. 2006;45(5):554–6.
19. Mat C, Bang D, Melikoğlu M. The mucocutaneous manifestations and pathergy reaction in Behçet's disease. In: Yazıcı Y, Yazıcı H, editors. Behçet's syndrome. 1st ed. New York: Springer; 2010.
20. Hatemi G, Fresko I, Tascilar K, Yazici H. Increased enthesopathy among Behçet's syndrome patients with acne andarthritis: an ultrasonography study. Arthritis Rheum. 2008;58(5):1539–45. https://doi.org/10.1002/art.23450.
21. Kuzu MA, Ozaslan C, Köksoy C, Gürler A, Tüzüner A. Vascular involvement in Behçet's disease: 8-year audit. World J Surg. 1994;18(6):948–53; discussion 953–4.
22. Alpsoy E, Zouboulis CC, Ehrlich GE. Mucocutaneous lesions of Behcet's disease. Yonsei Med J. 2007;48(4):573–85. Review.
23. Jorizzo JL, Abernethy JL, White WL, Mangelsdorf HC, Zouboulis CC, Sarica R, et al. Mucocutaneous criteria for the diagnosis of Behçet's disease: an analysis of clinicopathologic data from multiple international centers. J Am Acad Dermatol. 1995;32(6):968–76.
24. Yurdakul S, Hamuryudan V, Fresko I, Yazıcı H. Behçet's syndrome. In: Hochberg MC, Silman AJ, Smolen JS, Weinblatt WE, Weisman MH, editors. Rheumatology. 4th ed. Philadelphia: Mosby-Elsevier; 2008.
25. Yurdakul S, Yazici H, Tüzün Y, Pazarli H, Yalçin B, Altaç M, et al. The arthritis of Behçet's disease: a prospective study. Ann Rheum Dis. 1983;42(5):505–15.
26. Calamia KT, Schirmer M, Melikoglu M. Major vessel involvement in Behçet's disease: an update. Curr Opin Rheumatol. 2011;23:24–31.
27. Yazici H, Yurdakul S, Hamuryudan V. Behçet disease. Curr Opin Rheumatol. 2001;13:18–22.
28. Saadoun D, Wechsler B, Desseaux K, Le Thi Huong D, Amoura Z, Resche-Rigon M, et al. Mortality in Behcet's disease. Arthritis Rheum. 2010;62:2806–12.
29. Korman U, Cantasdemir M, Kurugoglu S, Mihmanli I, Soylu N, Hamuryudan V, et al. Enteroclysis findings of intestinal Behcet disease: a comparative study with Crohn disease. Abdom Imaging. 2003;28(3):308–12.

Neuro-Behçet Syndrome: Differential Diagnosis

Ugur Uygunoglu and Aksel Siva

Given that Behçet syndrome (BS) shares features with both autoinflammatory and autoimmune diseases, both groups of diseases affecting the central nervous system remain in the differential diagnosis of parenchymal neuro-Behçet syndrome (p-NBS).

However, the unique magnetic resonance imaging (MRI) pattern of p-NBS facilitates the diagnosis of p-NBS, particularly in regions where p-NBS is prevalent [1]. The prevalence of BS in Western countries has increased with immigration from countries where BS is prevalent; this has raised awareness of p-NBS. The differential diagnosis of BS is an important topic in these countries. Although the differential diagnosis of p-NBS includes many central nervous system diseases, this chapter addresses the most common diseases that might be confused with p-NBS [2]. These diseases are summarized in Table 1.

Four major features are used to differentiate p-NBS from other central nervous system diseases: clinical features, cerebrospinal fluid findings, pathological findings, and MRI patterns. As the clinical features of p-NBS were discussed in the previous section, they are not included in this chapter.

Differentiation of p-NBS from Other Mimickers by MRI Patterns

Cranial MRI and magnetic resonance venography (MRV) are of utmost importance for the diagnosis of NBS and for differentiating it from the other disorders mimicking NBS patterns in MRI. Although venous infarcts may occur rarely in cerebral venous sinus thrombosis (CVST), telencephalic lesions may happen due to venous infarcts.

U. Uygunoglu · A. Siva (✉)
Istanbul University, Cerrahpaşa School of Medicine, Department of Neurology, Istanbul, Turkey

E. Tüzün, M. Kürtüncü (eds.), *Neuro-Behçet's Disease*,
https://doi.org/10.1007/978-3-030-55273-2_7

Table 1 The differential diagnosis of intra-axial (CNS) neuro-Behçet's syndrome

Primary neurologic disorders
Multiple sclerosis
Stroke in young adults
CNS vasculitis
Neuro-sarcoidosis
CNS tuberculosis
Brainstem glioma
High-grade astrocytoma
Primary CNS lymphoma
Systemic disease with neurologic involvement
Secondary CNS vasculitis
Vogt-Koyanagi-Harada syndrome
Reiter syndrome
Eales' disease
Cogan's syndrome
Susac syndrome
Neuro-Sweet syndrome

Modified from Siva and Saip [2]

In p-NBS, cranial MRI shows an almost stereotypical lesions involving the brainstem, mainly the midbrain and upper pons, and extending to the diencephalon and basal ganglia, as well as a caudal extension in some. The lesions are hyperintense on T2 images and hypo-/iso-hypointense on T1 images; usually, there is a much smaller area of enhancement, and occasionally, small hemorrhages can be seen within the lesions. After steroid treatment, the lesions regress to punctate T2 hyperintense areas and brainstem atrophy develops [1].

When there is spinal cord involvement, it may tend to be longitudinally extensive [3, 4]. Although the long lesions are similar to those seen in patients with neuromyelitis optica spectrum disorder (NMOSD), BS patients do not fulfill the clinical criteria for NMOSD, and to the best of our knowledge, none of the BS patients with myelopathy have been reported as positive for the anti-aquaporin 4 antibody. Regarding the spinal cord involvement in BS, we described two distinct MRI patterns according to T2-weighted axial images: (a) "bagel sign" pattern, a central lesion with hypointense core and hyperintense rim with or without contrast enhancement, and (b) "motor neuron" pattern, a symmetric involvement of the anterior horn cells [5].

When considering the parenchymal distribution of lesions in NBS, lesions seem to support the hypothesis of small vessel vasculitis, mainly venular involvement. The known anatomic arrangement of CNS intra-axial veins explains the predominant involvement of the brainstem structures. This pattern of lesion distribution might help to differentiate NBS from other vasculitis as well as from the inflammatory-demyelinating diseases of the CNS, such as multiple sclerosis (MS) [1, 5, 6].

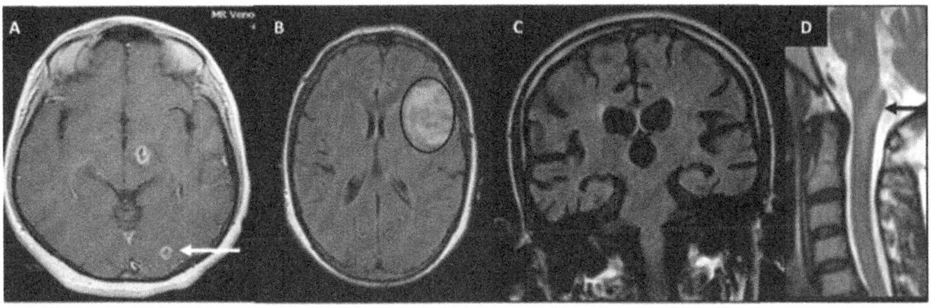

Fig. 1 Brain MRI of atypical parenchymal patterns during the parenchymal neuro-Behçet syndrome episodes are shown. (**a**) Axial T1 Gd (+) image shows a typical midbrain-diencephalic pattern of involvement for NBS but an additional atypical occipital lesion for NBS. (**b**) Axial FLAIR image shows a hemispheric tumefactive lesion. (**c**) Coronal FLAIR image reveals significant global cerebral atrophy. (**d**) MRI reveals area postrema lesion in a patient having BS for 10 years. AQP4 antibody was negative

Rarely, instead of the typical brainstem-diencephalic lesion/s, atypical patterns and involvement at other parenchymal sites may occur (Fig. 1) [7].

According to MRI features, we recently described four patterns:

1. Bilateral cortical-subcortical lesions
2. Hemispheric tumefactive lesions
3. Regional or global atrophy
4. Area postrema involvement

As the treatment options differ, NBS diagnosis should be carefully done in patients whose MRIs show lesions other than brainstem-diencephalon. In our clinical experience, we see many patients without any significant neurological complaint suggestive of NBS and fulfilling the BS criteria with an MRI showing bilateral cortical-subcortical lesions similar to lesions seen in MS. However, from the clinical point of view, MS is more common in women, whereas NBS is seen frequently in men, and optic neuritis, sensory symptoms, and spinal cord involvement, which are common in MS, are rarely seen in NBS. Nevertheless, in some cases, the clinical presentation of NBS may be confused with MS, but the neuroimaging (MRI) findings are clearly different. Together with the clinical differences and MRI features, NBS lesions could easily be differentiated from MS. Below are the main differences of MS lesions from NBS lesions [6].

1. The posterior fossa lesions of MS are small and discrete, whereas p-NBS brainstem lesions are large and diffuse, may have mass effect, and extend toward the diencephalic, thalamic, and basal ganglia regions and only rarely to the optic nerve regions.
2. Periventricular, juxtacortical, and corpus callosum lesions are common in MS, but rare in p-NBS.

3. Hemispheric subcortical regions are rare in p-NBS, and when present, they are usually small and asymptomatic, whereas coalescent periventricular lesions are more supportive of MS.
4. Spinal cord involvement rarely extends more than a few vertebral segments in MS.

Besides these differences, Maggi et al. recently published a paper comparing the frequency of perivenular lesions between MS and other inflammatory vasculopathies [8]. Although they found that the NBS patients revealed higher frequency of venular lesions from the other inflammatory vasculopathies, it was shown that a subgroup of BS patients with neurological complaints may have MS-like lesions in MRI, and MRI may even fulfill the radiological criteria of MS.

Brainstem lesions extending into the diencephalic region and basal ganglia during acute disease may exert mass effects caused by vasogenic edema and thus resemble tumors [9]. Brainstem glioma and primary CNS lymphoma should be included in the differential diagnosis of NBS in patients presenting with localized brainstem findings and whose initial MRI discloses a large brainstem lesion, but the presence of systemic findings, location of lesions other than the brainstem and deep hemispheric structures such as the frontoparietal or temporal lobe, and the resolution of the MRI lesion following high-dose steroids may solve the problem [10]. Although the inflammatory nature could not be shown in all cases, these lesions are likely to be acute inflammatory edematous lesions that following IV methylprednisolone (IVMP) show significant resolution. Brainstem and global atrophy may provide an important clue of a progressive NBS even if the patient does not fulfill the ISG criteria for BS. Cerebellar involvement may rarely happen at acute phase, but it is not uncommon during the progressive phase of NBS [11].

Herpes simplex lesions should be considered in the differential diagnosis of genital ulcers, and cytological or virological examination should be performed as necessary. Tuberculosis may resemble BS because of its similar multisystem involvement, including subacute or chronic meningitis. However, hilar lymphadenopathy and pulmonary cavities are rare in BS. Microbiological and pathological examination of body fluids or tissue specimens should be able to clarify whether the disease is tuberculosis. Some of the systemic features helpful in the differential diagnosis of BS are discussed in detail in a recent review by Yazici [12].

Sarcoidosis can be confused with Behçet disease due to uveitis, arthritis, central nervous system disease, and increased globulins. The rate of neurologic involvement is also around 5% to 10% in sarcoidosis, with facial nerve palsy, optic neuropathy, headache, and seizures being the most common symptoms. The clinical, cerebrospinal fluid, and imaging features of neurosarcoidosis may mimic NBS; however, meningeal involvement is extremely rare in NBS and when seen is more likely to be associated with neurosarcoidosis [13]. Hilar lymph nodes on chest X-ray, serum angiotensin-converting enzyme levels, Kveim skin test, and pathological examination of the granulomatous lesions of sarcoidosis help in the differential diagnosis.

An acute stroke-like onset is not common in NBS, and MRI lesions compatible with classical arterial territories are also not expected. The absence of systemic symptoms and signs will serve to differentiate the primary CNS vasculitic disorders from NBS and the difference in the systemic symptoms and signs from the secondary CNS vasculitides, as well as the MRI findings [14]. The inflammatory large brainstem lesions occasionally extending to the deep hemispheric and diencephalic regions seen in most acute CNS-NBS patients are unlikely to be seen in other systemic vasculitic disorders.

In primary CNS vasculitis, cerebral angiography is reported to be abnormal in up to 90% of patients, and MRI can detect multiple infarcts, mostly involving cortical areas [15], which are unusual in NBS. There is no systemic involvement, and the MRI lesions in primary CNS vasculitis are more likely to be hemispheric with subcortical and cortical or juxtacortical involvement, not common in NBS [14].

Due to their ophthalmologic and some other systemic manifestations, rare diseases, such as Vogt-Koyanagi-Harada syndrome, Reiter syndrome, Eales' disease, Cogan's syndrome, and Susac syndrome, are other considerations in the differential diagnosis of BS [2]. All may present with nervous system manifestations and therefore are also included in the differential diagnosis of NBS. However, a complete ophthalmologic examination will reveal the true nature of eye involvement in each of these syndromes, which have differences from the eye involvement seen in BS.

In patients who present with symptoms of intracranial hypertension and in whom neuroimaging reveals thrombosis in one or more of the cerebral venous sinuses, BS needs to be included in the differential diagnosis. The presence of its systemic findings is the only clue to the association of CVST with BS, and their absence will exclude this possibility. One of the main difference of CVST due to BS from other etiologies causing CVST is that hemorrhagic venous infarcts are rarely seen in BS-CVST [16].

Differentiation of p-NBS from Other Mimickers by Laboratory Findings

During the acute phase of p-NBS, the CSF shows inflammatory changes in most cases of p-NBS with an increased number of cells, up to a hundred and sometimes more per ml, neutrophils being mostly the predominating cells and modestly elevated protein levels. However, an early lymphocytic pleocytosis is not an exception. When the neutrophilic pleocytosis is the case, it is later replaced by lymphocytes. As stated above, prominent inflammatory changes with CSF pleocytosis and mild to moderate elevation of protein in the CSF are more common in p-NBS than in multiple sclerosis. Oligoclonal bands are found in more than 90% of multiple sclerosis patients and in less than 20% of NBS patients [17].

As already noted, patients with BS-CVST do not exhibit any remarkable CSF finding apart from an increased pressure.

An elevated concentration of IL-6 in the CSF of patients correlating with disease activity has also been reported in p-NBS [18–20]. More recently, an increase of CSF IL-10 and CSF/serum matrix metallopeptidase-9 ratio (increase in serum and decrease in CSF compared to multiple sclerosis) was reported and suggested to be a discriminative marker between NBS and multiple sclerosis [21, 22].

Differentiation of p-NBS from Other Mimickers by Pathological Findings

The main pathological differences between multiple sclerosis (MS) and NBS were described in 1957. Still today, MS is considered in the differential diagnosis of NBS [23]. The earlier necropsies described precisely what we observe in clinical practice and pathology specimens:

1. NBS does not feature an extensive and varied pattern of demyelination (unlike MS).
2. NBS lacks intense fibrous gliosis.
3. Cord lesions tend to be peripheral in MS but more central in BS.
4. Periventricular lesions are expected in MS, but not in BS.

When considering all published autopsy cases together with the clinical and MRI findings, the pathogenesis of NBS may be vasculitis, or combined perivasculitis/capillaritis, accompanied by lymphocytic infiltration, hemorrhage, demyelination, axon destruction, oligodendroglial degeneration, migration of fatty granule cells, proliferation of activated microglia, occasional glial nodules, and gliomesenchymal fibrosis [24].

References

1. Kocer N, Islak C, Siva A, et al. CNS involvement in neuro-Behcet's syndrome: an MR study. Am J Neuroradiol. 1999;20:1015–24.
2. Siva A, Saip S. The spectrum of nervous system involvement in Behcet's syndrome and its differential diagnosis. J Neurol. 2009;256:513–29.
3. Lee HS, Kim do Y, Shin HY, et al. Spinal cord involvement in Behcet's disease. Mult Scler. 2016;22:960–3.
4. Uygunoglu U, Pasha M, Saip S, Siva A. Recurrent longitudinal extensive transverse myelitis in a neuro-Behcet syndrome treated with infliximab. J Spinal Cord Med. 2015;38:111–4.
5. Uygunoglu U, Zeydan B, Ozguler Y, et al. Myelopathy in Behcet's disease: the bagel sign. Ann Neurol. 2017;82:288–98.
6. Siva A. Common clinical and imaging conditions misdiagnosed as multiple sclerosis: a current approach to the differential diagnosis of multiple sclerosis. Neurol Clin. 2018;36:69–117.
7. Siva A, Esatoglu SN, Uygunoglu U, et al. Could neurological involvement in Behcet's disease occur atypically? (P5.410). Neurology. 2018;90(15 Supplement).

8. Maggi P, Absinta M, Grammatico M, et al. Central vein sign differentiates multiple sclerosis from central nervous system inflammatory vasculopathies. Ann Neurol. 2018;83:283–94.
9. Cohen-Aubart F, Psimaras D, Galanaud D, et al. Cerebral pseudo-tumoral neuro-Behcet: histological demonstration of an inflammatory and vascular disease. Clin Neurol Neurosurg. 2017;161:48–50.
10. Imataki O, Uchida S, Yokokura S, Uemura M, Kadowaki N. Central nervous system peripheral T cell lymphoma manifesting as Lymphomatosis cerebri that was misdiagnosed as Neuro-Behçet's disease: a case report. Case Rep Oncol. 2018;11(3):806–13.
11. Taskapilioglu O, Seferoglu M, Akkaya C, et al. Delayed diagnosis of a neuroBehcet patient with only brainstem and cerebellar atrophy: literature review. J Neurol Sci. 2009;277:160–3.
12. Yazici H, Seyahi E, Hatemi G, Yazici Y. Behcet syndrome: a contemporary view. Nat Rev Rheumatol. 2018;14:107–19.
13. Uygunoglu U, Saip S, Siva A. Chapter 28: Behcet's syndrome and nervous system involvement. In: Lisak RP, Truong DD, Carroll WM, et al., editors. International neurology. 2nd ed. Chichester/Hoboken: John Wiley & Sons, Ltd; 2016. p. 88–93.
14. Uygunoğlu U, Siva A. Behçet's syndrome and nervous system involvement. Curr Neurol Neurosci Rep. 2018;18(7):35.
15. Salvarani C, Brown RD Jr, Calamia KT, et al. Primary central nervous system vasculitis: analysis of 101 patients. Ann Neurol. 2007;62(5):442–51.
16. Yesilot N, Bahar S, Yilmazer S, et al. Cerebral venous thrombosis in Behcet's disease compared to those associated with other etiologies. J Neurol. 2009;256:1134–42.
17. Saruhan-Direskeneli G, Yentür SP, Mutlu M, Shugaiv E, Yesilot N, Kürtüncü M, et al. Intrathecal oligoclonal IgG bands are infrequently found in neuro-Behçet's disease. Clin Exp Rheumatol. 2013;31(3 Suppl 77):25–7.
18. Hirohata S, Isshi K, Oguchi H, et al. Cerebrospinal fluid interleukin-6 in progressive neuro-Behçet's syndrome. Clin Immunol Immunopathol. 1997;82(1):12–7.
19. Akman-Demir G, Tuzun E, Icoz S, et al. Interleukin-6 in neuro-Behçet's disease: association with disease subsets and long-term outcome. Cytokine. 2008b;44(3):373–6.
20. Hirohata S, Kikuchi H. Changes in biomarkers focused on differences in disease course or treatment in patients with neuro-Behçet's disease. Intern Med. 2012a;51(24):3359–65.
21. Aldinucci A, Bonechi E, Biagioli T, et al. CSF/serum matrix metallopeptidase-9 ratio discriminates neuro Behçet from multiple sclerosis. Ann Clin Transl Neurol. 2018;5(4):493–8.
22. Belghith M, Bahrini K, Kchaou M, Maghrebi O, Belal S, Barbouche MR. Cerebrospinal fluid IL-10 as an early stage discriminative marker between multiple sclerosis and neuro-Behçet disease. Cytokine. 2018;108:160–7.
23. Berlin C. Behçet syndrome with involvement of the central nervous system. Arch Derm Syphilol. 1944;49:227–33.
24. McMenemey WH, Lawrence BJ. Encephalomyelopathy in Behçet disease: report of necropsy findings in two cases. Lancet. 1957;2:353–8.

Neuro-Behçet Syndrome: Clinical Features

Ugur Uygunoglu and Aksel Siva

Given that there is no existing diagnostic biomarker available, the diagnosis of neuro-Behçet syndrome (NBS) relies on symptoms, signs, and radiological features suggestive of NBS. Therefore, clinical features and neurological examination play a major role in the diagnosis of NBS. In the meantime, NBS is not included in the current criteria of International Study Group (ISG). However, considering the rate of 5–10% neurologic involvement among BS patients and taking into account the increased atypical presentation of neurological involvement particularly over the last years, the importance of NBS is increasing [1]. The frequency of neurologic involvement was found to reach 13% in males and 5.6% in females when patients were followed longitudinally for up to 20 years [2]. NBS generally presents in individuals in their late 30s, is more common in males (3:1), and is characterized by an interval of approximately 5 years between BS onset and the first neurological episode [3]. Up to 6% of patients may present with neurological involvement without fulfilling the ISG classification criteria for BS, which is the most challenging factor for clinicians in terms of making an accurate diagnosis and deciding when to initiate long-term treatment.

In 2014, the clinical, laboratory, and neuroimaging features of "definite" and "probable" NBS were established by a panel of NBS experts [4]. The current criteria are slight modifications of the "Cerrahpaşa-NBS criteria" introduced in 2001 [5]. These can be summarized as: "The occurrence of neurological symptoms and signs in a patient who meets the International Study Group Criteria for BD, when those symptoms/signs are not otherwise referable to any other known systemic or neurological disease, or treatment thereof, and in whom objective abnormalities consistent with NBS are evident either on neurological examination or MRI, or upon cerebrospinal fluid (CSF) analysis." The principal novelty of the new criteria is that "probable NBS" is now identifiable by consensus. Probable NBS is defined as either of the following conditions when the neurological findings do not indicate

U. Uygunoglu (✉) · A. Siva
Istanbul University Cerrahpaşa School of Medicine, Department of Neurology, Istanbul, Turkey

E. Tüzün, M. Kürtüncü (eds.), *Neuro-Behçet's Disease*,
https://doi.org/10.1007/978-3-030-55273-2_8

another neurological disease: (1) neurological syndrome(s) suggestive of definitive NBS, but systemic features that do not meet the International Study Group (ISG) BS criteria, or (2) a non-characteristic neurological syndrome within the context of BS supported by the ISG criteria. However, neither the International Consensus Recommendation (ICR) nor the Cerrahpaşa-NBS criteria have yet been validated. Given the increased use of new biological agents and the various comorbidities of BS, we more often encounter patients with "probable" than with "definite" NBS [6].

Neurological involvement in BS can be classified into two forms:

1. Primary
2. Secondary

The primary neurological conditions include p-NBS, cerebral venous sinus thrombosis (CVST), neuro-psycho-Behçet syndrome, cognitive changes, headache (migraine-like, nonstructural), peripheral nervous system involvement, and subclinical NBS (Table 1). However, by reference to clinical/neuroimaging features, primary BS neurological involvement can be classified into two principal forms:

1. p-NBS
2. Neurovascular involvement

Table 1 The neurologic spectrum of Behçet syndrome

Primary neurologic involvement (neurologic involvement directly related to BS)
Neurovascular BS – involvement of the extra-parenchymal vessels
Cerebral venous sinus thrombosis (extra-axial NBS)
Large arterial involvement – i.e., arterial dissection, aneurysms
Parenchymal NBS – central nervous system involvement
Post-capillary venular involvement
Involvement of parenchymal small arteries
Neuro-psycho-Behçet syndrome (NPBS)
With CNS parenchymal disease
Without CNS parenchymal disease
Cognitive changes
Isolated headache syndrome (migraine-like, nonstructural)
Peripheral nervous system involvement
Subclinical NBS
Secondary neurologic involvement (neurologic involvement indirectly related to BS)
Neurologic complications secondary to systemic involvement of BS (i.e., cerebral emboli from cardiac complications of BS, increased intracranial pressure secondary to superior vena cava syndrome)
Neurologic complications related to BS treatments (i.e., CNS neurotoxicity with cyclosporine, peripheral neuropathy secondary to thalidomide or colchicine)
Somatoform neurologic symptoms associated with having a chronic disease
Coincidental – unrelated (non-BS) neurologic involvement
Primary headaches – as in the general population
Any coincidental neurologic disorders

Modified and updated from Uygunoglu and Siva [1]
BS Behçet syndrome, *NBS* neuro-Behçet syndrome, *CNS* central nervous system

Parenchymal NBS

Parenchymal NBS (p-NBS) is one of the most disabling complications of BS and develops in approximately 75–80% of adult NBS cases. Presentation is usually either acute or subacute; however, in some cases, superimposed exacerbations during progression may be observed [7]. Considering the subacute progressive presentation of the clinical findings, the distribution of the p-NBS lesions prominently in the area where the venous anastomosis is few, the significant resolution of the perilesional edema with some small residua, and the pathological findings regarding p-NBS, all these features support the venous pathogenesis in the development of p-NBS [7–9].

Clinically, neurologic manifestations are usually related to brainstem or corticospinal tract syndromes, with the most common presentation being a subacute brainstem syndrome that includes ophthalmoparesis and other cranial nerve findings, dysarthria, and corticospinal tract signs with or without weakness and ataxia. The presentation may include all or some of these symptoms and signs, and during the acute stage, a mild confusion may also be seen [10].

It should be kept in mind that parenchymal NBS (intra-axial NBS) does not always present with brainstem signs and symptoms. Cognitive-behavioral changes, emotional lability, a self-limited or progressive myelopathy, urinary sphincter dysfunction, and to a lesser extent other CNS manifestations such as extrapyramidal signs and seizures have been reported. There are also a few cases with isolated optic neuritis or recurrent peripheral facial paresis [11].

Isolated progressive ataxia with cerebellar atrophy on MRI has been reported in a few patients with BS, and it was suggested that this form of presentation may be novel manifestation of NBS [12]. However, since the relationship between the neurological presentation and BS was not clear in those cases and comorbidity could not be ruled out, further observation is needed before such a conclusion may be reached.

Aseptic meningitis was reported previously to be a relatively frequent form of neurological involvement in patients with BS, in studies, which was based on CT scans as the primary imaging modality [13]. Our experience has not been such, and we suspect that in some patients, parenchymal disease was misclassified as aseptic meningitis due to lack of sensitive imaging data. As a matter of fact, in a more recent update from the same institution [14], aseptic meningitis was reported only in one out of 200 cases studied. Similarly, in another study of 50 patients from the UK [15], four cases were reported to have meningitis symptoms, while two of these patients had parenchymal lesions and two had normal MRI. There was no discussion of meningeal enhancement, and the CSF findings were within the same range as with patients who had brainstem parenchymal involvement. In support of this, we have always observed inflammatory findings in CSF together with parenchymal disease in MRI. Taken together, we conclude that pure aseptic meningitis is very rare within the clinical spectrum of neurological involvement in BS [5].

Regarding the clinical phenotype p-NBS, it may be classified into four subtypes similar to multiple sclerosis (MS): (1) single attack, (2) relapsing form, (3) secondary progressive, and (4) primary progressive. The onset of p-NBS is usually associated with flaring of systemic features of BS. Some patients may only have a single attack, with or without residual neurologic deficits, but most will have recurrences with further sequelae, and some will have secondary progression. A small number of patients may have a primary progressive course [16]. A recent meta-analysis of a Japanese cohort which investigated the distinct clinical features of the acute and chronic phases of the progressive forms of p-NBS reported that acute p-NBS patients present with more brain stem symptoms, are more likely to have meningoencephalitis with high fever, and have elevated CSF cell counts [17]. On the other hand, confusion, dementia, dysarthria, and ataxia were found to be more common in the chronic progressive phase of p-NBS.

In terms of clinical outcome, although no outcome measures have been validated, we use Kurtzke's EDSS to determine the disability in NBS, which was originally devised for MS-associated disability [18]. We exclude the visual function score to avoid the contribution of uveitis to the visual score [5]. The EDSS is an ordinal clinical rating scale ranging from 0 (normal neurologic examination) to 10 (death due to MS) in half-point increments. From the perspective of clinical disability, severe relapses having high EDSSs, frequent relapses, and early disease progression are the poor prognostic features for NBS. Initiation with severe disability, primary or secondary progressive course, fever at onset, relapse during steroid tapering, meningeal signs, and bladder involvement are possible association with poor outcome. Initiation with severe disability, primary or secondary progressive course, fever at onset, relapse during steroid tapering, meningeal signs, and bladder involvement are possible association with poor outcome whereas gender, accompanying systemic features, and age of onset do not have impact on the prognosis of NBS [3, 5].

Neurovascular Involvement

BS is classified as a variable vessel vasculitis [19]. Of the 251 autopsies described from 1961 to 1987, 18% were classified as having vascular BS, including whole-system vasculitis [20]. In terms of the neurovascular structures involved, two major types of vascular pathology are seen in BS patients:

(a) Venous thrombosis
(b) Arterial occlusion and/or aneurysm formation

Cerebral Venous Sinus Thrombosis (CVST)

Cerebral venous sinus thrombosis (CVST) occurs in up to 20% of BS patients with neurological involvement. In such patients, the principal clinical features (severe headache, papilledema, and sixth nerve palsy on neurological examination) are compatible with intracranial hypertension [5]. The overall incidence of CVST in

3908 patients with BS was 3.1/1000 person-years. Notably, 30% did not fulfill the ISG criteria either at CVST commencement or thereafter; this proportion is much higher than that of suspected p-NBS patients [21]. Given that most studies show that BS-associated CVST has a good prognosis in contrast to other etiologies causing CVST, the systemic features of BS in CVST patients, especially those living in endemic regions, should be assessed for BS. CVST is usually subacute or chronic; only about 25% of cases exhibit clinical features for more than 1 month [22]. Such longer presentation may reflect the time frame of thrombosis and may help clinicians to distinguish BS-CVST from other causes of CVST. Hemiparesis, impaired consciousness, and epileptic seizures are uncommon in CVST patients with NBS which may be explained by the extremely low probability of seeing hemorrhagic venous infarcts associated with NBS-CVST [21–23]. It can be speculated that the CVST in BS develops gradually and allows sample time for the development of efficient collaterals and therefore does not cause manifestations that may be seen with acute onset CVST due to other causes. Cranial MRI and magnetic resonance venography (MRV) will show that the most commonly involved dural venous sinuses are the superior and transverse sinuses, followed by the sigmoid and straight sinuses. Single-sinus occlusion is more frequent than multiple occlusions [21, 22]. Additionally, clinicians should be aware that cranial MRI and MRV scans may not show sinus thrombosis, even if the clinical findings strongly suggest its presence. In such situations, MRV of the thoracic and cervical venous structures should be evaluated. Irrespective of whether the neuroimaging data are abnormal or normal, we generally perform a spinal tap to study CSF pressure and contents in the suspected cases. Apart from CVST, extracranial thrombosis of large vessels is observed in half of all BS patients, whereas the prevalence of extracranial vascular involvement is about 20% in BS patients without CVST [21]. Therefore, BS patients with CVST should undergo evaluation for vascular involvement at other locations. Two case series found that CVST was more common in younger patients, supporting the idea that age is important in terms of NBS presentation [24, 25]. Although an elevated opening pressure is observed, interestingly, the CSF is free of inflammatory changes in BS-CVST patients.

Arterial Occlusion and Aneurysm Formation

Arterial involvement affects only 3–5% of BS patients but is unique to BS; aneurysms may occur in the peripheral, visceral, and pulmonary arteries. As the aneurysms are probably attributable to chronic inflammation, vascular involvement may be evident in the later phases of the syndrome. The prevalence is probably underestimated as autopsy data reveals a significantly higher rate of such arterial lesions than the clinically reported ones [20]. Saccular and multiple dissecting aneurysms predominated in autopsy series. The cellular infiltrates were predominantly neutrophils, lymphocytes, and plasma cells, admixed with histiocytes and eosinophils. Intimal thickening of the vasa vasorum was also observed. During the acute phase, inflammatory cells were more common in the media and adventitia than in the intima, principally in the proliferating vasa vasorum. Such intense inflammation triggers destruction of the media and development of saccular aneurysms [26]. In addition, bilateral internal carotid artery occlusion, vertebral artery dissection or thrombosis, intracranial arteritis, and intra-axial small arterial occlusion have been reported in BS patients [27].

Primary Neurological Conditions Other than p-NBS and Neurovascular Involvement

(a) *Neuro-psycho-Behçet syndrome*: A neurobehavioral syndrome, which may be expressed with a number of symptoms such as euphoria, loss of insight, disinhibition, indifference to the disease, psychomotor agitation or retardation with paranoid attitudes, and obsessive concerns, may be seen in a minority of patients with BS. This syndrome that is unrelated to glucocorticoid or any other therapy is named neuro-psycho-Behçet syndrome [28]. This neurobehavioral syndrome may be seen in association with p-NBS but at times may be isolated without any structural change.

(b) *Cognitive changes*: Cognitive changes consistent with a combination of memory impairment in the form of delayed recall, abnormalities in verbal and/or visual modalities, and impairment in the process of acquisition and storage, in addition to attention deficit and deficits of executive function of the frontal system, have all been reported in BS [29, 30]. These cognitive changes may be observed in patients with BS either unrelated to other forms of neurological involvement or in association with p-NBS.

(c) *Headache*: Headache is the most common neurologic symptom seen in BS and can be due to various causes. It may be the presenting symptom of either form of NBS; it can be seen as a symptom of ocular inflammation or may be independent of the disease simply being a primary headache of the migraine or tension type [1]. Up to 20% of people with BS may report a bilateral, frontal, moderately severe paroxysmal migraine-like throbbing pain, which is not true migraine and does not uncommonly accompany the exacerbations of systemic findings of BS such as oral ulcerations or skin lesions [31–33]. This headache that is not related to either p-NBS or CVST is likely to be a toxic-vascular headache triggered by the immune-mediated disease activity in susceptible individuals [1]. However, BS patients who report a severe headache of recent onset in the absence of any neurologic deficit and not consistent with any primary headache should be evaluated carefully to rule out the onset of NBS.

(d) *Peripheral nervous system (PNS) involvement*: PNS disease with clinical manifestations is extremely rare in BS. Mononeuritis multiplex, polyradiculoneuritis, sensorimotor axonal neuropathy, or recurrent episodes of myositis had been diagnosed by the clinical manifestations and electrophysiological findings in a handful cases with BS. Probably, more cases of neuropathy as a result of thalidomide or colchicine use have been observed [1].

(e) *Subclinical NBS*: The incidental finding of neurologic signs in patients with BS without neurologic symptoms has been reported in some series, with a minority of these patients developing mild neurologic attacks in due course [10]. The presence of cognitive dysfunction that is only apparent on testing is also suggestive of subclinical neurologic involvement [30]. Abnormalities detected by a number of diagnostic tools such as MRI, SPECT and somatosensory evoked potentials in some BS patients who did not have any neurologic symptoms and signs were also noted [34].

Neurological complications secondary to systemic involvement of BS, such as cerebral emboli due to cardiac complications or increased intracranial pressure secondary to superior vena cava syndrome, are indirect causes of neurological problems in BS patients but are not generally termed NBS [35].

Regarding Clinical Features, When Should We Suspect NBS?

The onset of a subacute brainstem syndrome in a young man, especially of Mediterranean, Middle Eastern, or Oriental origin, that includes cranial nerve findings, dysarthria, unilateral or bilateral corticospinal tract signs with or without weakness, ataxia, and mild confusion should raise the possibility of NBS. The patient or his family should be interviewed regarding the presence of systemic findings of BS. In BS, there is likely a history of oral aphthous ulcers and other systemic manifestations of the disease.

Many patients may have never consulted a physician because of the mild nature of their systemic symptoms or may be missed because they do not develop a full-blown picture of the disease.

References

1. Uygunoğlu U, Siva A. Behçet's syndrome and nervous system involvement. Curr Neurol Neurosci Rep. 2018;18(7):35.
2. Kural-Seyahi E, Fresko I, Seyahi N, et al. The long-term mortality and morbidity of Behcet syndrome: a 2-decade outcome survey of 387 patients followed at a dedicated center. Medicine (Baltimore). 2003;82:60–76.
3. Al-Araji A, Kidd DP. Neuro-Behçet's disease: epidemiology, clinical characteristics, and management. Lancet Neurol. 2009;8:192–204.
4. Kalra S, Silman A, Akman-Demir G, et al. Diagnosis and management of Neuro-Behçet's disease: international consensus recommendations. J Neurol. 2014;261(9):1662–76.
5. Siva A, Kantarci OH, Saip S, et al. Behçet's disease: diagnostic & prognostic aspects of neurological involvement. J Neurol. 2001;248:95–103. 24.

6. Siva A, Esatoglu SN, Uygunoglu U, et al. Could neurological involvement in Behçet's disease occur atypically? Neurology. 2018;90(15 Supplement):P5. 410.
7. Uygunoglu U, Zeydan B, Ozguler Y, et al. Myelopathy in Behçet's disease: the bagel sign. Ann Neurol. 2017;82(2):288–98.
8. Kocer N, Islak C, Siva A, et al. CNS involvement in neuro-Behcet's syndrome: an MR study. Am J Neuroradiol. 1999;20:1015–24.
9. McMenemey WH, Lawrence BJ. Encephalomyelopathy in Behçet disease: report of necropsy findings in two cases. Lancet. 1957;2:353–8.
10. Siva A, Saip S. The spectrum of nervous system involvement in Behçet's syndrome and its differential diagnosis. J Neurol. 2009;256(4):513–29.
11. Uygunoglu U, Saip S, Siva A. Chapter 28: Behcet's syndrome and nervous system involvement. In: Lisak RP, Truong DD, Carroll WM, et al., editors. International neurology. 2nd ed. Chichester/Hoboken: John Wiley & Sons, Ltd; 2016. p. 88–93.
12. Taskapilioglu O, Seferoglu M, Akkaya C, et al. Delayed diagnosis of a neuroBehcet patient with only brainstem and cerebellar atrophy: literature review. J Neurol Sci. 2009;277:160–3.
13. Serdaroğlu P, Yazici H, Ozdemir C, Yurdakul S, Bahar S, Aktin E. Neurologic involvement in Behçet's syndrome. A prospective study. Arch Neurol. 1989;46(3):265–9.
14. Akman-Demir G, Serdaroglu P, Tasçi B. Clinical patterns of neurological involvement in Behçet's disease: evaluation of 200 patients. The Neuro-Behçet Study Group. Brain. 1999;122(Pt 11):2171–82.
15. Kidd D, Steuer A, Denman AM, Rudge P. Neurological complications in Behçet's syndrome. Brain. 1999;122(Pt 11):2183–94.
16. Siva A. Common clinical and imaging conditions misdiagnosed as multiple sclerosis: a current approach to the differential diagnosis of multiple sclerosis. Neurol Clin. 2018;36(1):69–117.
17. Ishido M, Horita N, Takeuchi M, Shibuya E, Yamane T, Kawagoe T, et al. Distinct clinical features between acute and chronic progressive parenchymal neuro-Behçet disease: meta-analysis. Sci Rep. 2017;7(1):10196.
18. Kurtzke JF. Rating neurologic impairment in multiple sclerosis: an expanded disability status scale (EDSS). Neurology. 1983;33:1444–52.
19. Jennette JC, Falk RJ, Bacon PA, et al. 2012 revised international Chapel Hill consensus conference nomenclature of vasculitides. Arthritis Rheum. 2013;65(1):1–11.
20. Lakhanpal S, Tani K, Lie JT, Katoh K, Ishigatsubo Y, Ohokubo T. Pathologic features of Behçet's syndrome: a review of Japanese autopsy registry data. Hum Pathol. 1985;16(8):790–5.
21. Aguiar de Sousa D, Mestre T, Ferro JM. Cerebral venous thrombosis in Behçet's disease: a systematic review. J Neurol. 2011;258(5):719–27.
22. Uluduz D, Midi I, Duman T, et al. Behçet's disease as a causative factor of cerebral venous sinus thrombosis: subgroup analysis of data from the VENOST study. Rheumatology (Oxford). 2018;58(4):600–8. https://doi.org/10.1093/rheumatology/key153.
23. Yesilot N, Bahar S, Yilmazer S, et al. Cerebral venous thrombosis in Behçet's disease compared to those associated with other etiologies. J Neurol. 2009;256(7):1134–42.
24. Uluduz D, Kürtüncü M, Yapıcı Z, et al. Clinical characteristics of pediatric-onset neuro-Behçet disease. Neurology. 2011;77(21):1900–5.
25. Shi J, Huang X, Li G, Wang L, Liu J, Xu Y, Zeng X, Zheng W. Cerebral venous sinus thrombosis in Behçet's disease: a retrospective case-control study. Clin Rheumatol. 2018;37(1):51–7.
26. Matsumoto T, Uekusa T, Fukuda Y. Vasculo-Behçet's disease: a pathologic study of eight cases. Hum Pathol. 1991;22(1):45–51.
27. Kizilkilic O, Albayram S, Adaletli I, et al. Endovascular treatment of Behçet's disease-associated intracranial aneurysms: report of two cases and review of the literature. Neuroradiology. 2003;45(5):328–34.
28. Siva A, Özdogan H, Yazici H, et al. Headache, neuro-psychiatric and computerized tomography findings in Behçet's syndrome. In: Lehner T, Barnes CG, editors. Recent advances in Behçet's disease. London: Royal Society of Medicine Service; 1986. p. 247–54.

29. Oktem-Tanor O, Baykan-Kurt B, Gurvit IH, Akman-Demir G, Serdaroglu P. Noel neuropsychological follow-up of 12 patients with neuro-Behcet disease. J Neurol. 1999;246(2):113–9.
30. Cavaco S, da Silva AM, Pinto P, et al. Cognitive functioning in Behçet's disease. Ann N Y Acad Sci. 2009;1173:217–26.
31. Saip S, Siva A, Altintas A, et al. Headache in Behcet's syndrome. Headache. 2005;45:911–9.
32. Borhani Haghighi A, Aflaki E, Ketabchi L. The prevalence and characteristics of different types of headache in patients with Behçet's disease, a case-control study. Headache. 2008;48(3):424–9.
33. Kale N, Agaoglu J, Icen M, Yazici I, Tanik O. The presentation of headache in neuro-Behçet's disease: a case-series. Headache. 2009;49(3):467–70.
34. Avci O, Kutluay E, Argon M, Erdem S, Tahsin Gunes A. Subclinical cerebral involvement in Behcet's disease: a SPECT study. Eur J Neurol. 1998;5(1):49–53.
35. Saip S, Akman-Demir G, Siva A. Neuro-Behçet syndrome. Handb Clin Neurol. 2014;121:1703–23.

The Cerebrospinal Fluid Presentations of Neuro-Behçet Disease

Afshin Borhani-Haghighi

As mentioned in previous chapters, the central nervous system (CNS) manifestations of Behçet disease can be categorized into two main groups:

1. Parenchymal CNS involvement of NBD: the more common type of presentation, which includes brainstem involvement, hemispheral manifestations, spinal cord lesions, and meningoencephalitic presentations.
2. Non-parenchymal CNS involvement (neurovasculo-Behçet disease) which includes venous and arterial involvement. The simultaneous parenchymal and non-parenchymal manifestations are rare [1, 2].

CSF laboratory changes in NBD are dependent on NBD pathogenesis and pathology. Vasculitis is the most probable pathologic presentation of Behçet disease. Venules are more affected than the arterioles, but large arteries and cerebral veins and sinuses are affected in some patients, as well. Perivascular infiltration of Polymorphonuclear (PMN) cells, mononuclear cells, and rarely eosinophilic cells has been seen in most pathological reports [3].

CSF Cells

In Nakamura et al.'s series, the mean CSF cell count was 126/mm^3 in the patients with NBD. The average percentages of mononuclear and polymorphonuclear cells were 67% and 32%, respectively [4].

In Al-Araji et al.'s study, in the patients with parenchymal NBD, the CSF opening pressure was within normal limits, but in pleocytosis with or without raised protein, raised protein was only observed in half of the CSF samples. Pleocytoses were PMN dominant and lymphocyte dominant equally. The sugar content was

A. Borhani-Haghighi (✉)
Clinical Neurology Research Center, Shiraz University of Medical Sciences, Shiraz, Iran

E. Tüzün, M. Kürtüncü (eds.), *Neuro-Behçet's Disease*,
https://doi.org/10.1007/978-3-030-55273-2_9

normal in all CSF samples. The patients with intracranial hypertension had increased CSF opening pressure but normal CSF cytochemistry [5].

In Akman-Demir et al.'s study, CSF analysis was performed in 81 patients with parenchymal NBD. CSF samples were completely normal. Forty-nine patients showed pleocytosis and/or elevated protein. Of these CSFs, 54% had either neutrophilic predominance or both neutrophilic and lymphocytic predominance, and 46% showed lymphocytic predominance.

In Kidd et al.'s series, 20 patients with active disease showed a mixed lymphocytic and neutrophilic pleocytosis, and there was no evidence for intrathecal synthesis of immunoglobulins.

In Borhani-Haghighi et al.'s study, CSF lymphocytes were between 6 and 30/μL (median: 18); CSF PMN was between 1 and 10/μL (median: 5).

CSF pleocytosis was seen in several studies [5–7].

The CSF cell count was prominently elevated in patients with acute NBD, but it was normal in about 15% of those with CP NBD. The sensitivity and specificity of the CSF cell count for the diagnosis of acute NBD versus non-NBD were 97.4 and 97.0%, respectively (cut-off: 6.2/mm^3) [8].

CSF Protein

In Hirohata's study, CSF total cell counts and polymorph nuclear leukocyte counts were significantly lower in patients with chronic progressive NB than in those with acute NB [9].

In Miller et al.'s case report, CSF analysis revealed elevated protein at 165 mg/dL [10].

In Borhani-Haghighi et al.'s study, CSF protein was between 41 and 96 mg/dL (median: 75).

In Al-Araji et al.'s study, increased CSF protein content was seen with or without pleocytosis in parenchymal NBD [5].

CSF Glucose

In Borhani-Haghighi et al.'s investigation, CSF glucose was between 51 and 79 mg/dL (median: 65).

Oligoclonal Bands

In Hirohata et al.'s (1986) study, CSF IgM index was significantly elevated in patients with active neuro-Behçet disease. Moreover, the CSF IgM index was found to decrease remarkably when the patients' neurologic manifestations disappeared after successful treatment [11].

In 1991, in Sharief et al.'s study, oligoclonal IgA and IgM bands were mainly detected in CSF samples from patients with active neuro-Behçet disease, and they were documented to disappear when neurologic manifestations remitted. Oligoclonal IgG bands, however, were not related to disease activity, and they were also found in some neurologic controls. High immunoglobulin index values were detected in both active and quiescent diseases and were high in some patients with impaired blood-CSF barriers [12].

In 1990, Gille et al. reported that immunoglobulin bands disappeared from CSF following an acute attack [13].

In 1992, Jongen et al. in a single case found that IgM and IgA bands disappeared following recovery from the episodes of meningoencephalitis [14].

In Mclean et al.'s study, local synthesis of oligoclonal IgG was found in 8% of NBD patients at some stages in their diseases by using isoelectric focusing and immunoblotting techniques. Blood-brain barrier breakdown when assessed with an albumin ratio was found in 42% of NBD patients. Reversibility of oligoclonal IgG, which is not the case in multiple sclerosis, had also been reported [15].

In Daoudi et al.'s study, oligoclonal bands were seen present in CSF of seven out of 40 patients with parenchymal NBD [16].

Oligoclonal bands disappeared or immunoglobulin indices decreased after acute attacks. Intrathecal 2-microglobulin synthesis has also been reported in NBD patients. In Saruhan-Direskeneli's study, only eight out of 121 patients with NBD showed positive oligoclonal bands (OCBs). All these positive cases had parenchymal neuro-BD [17].

In Miller et al.'s case report, IgG index and oligoclonal banding were all normal [18]. In Liu et al.'s case series with spinal NBD, CSF test was performed in 14 patients, indicating that the cell numbers and/or protein level were increased. Nine out of 14 CSF samples were tested by oligoclonal band and were totally negative [19].

Interleukin-6

In 1992, Wang et al. revealed that IL-6 was highly increased in the CSF of patients with NBD and decreased with clinical improvement [20].

In Hirohata et al.'s study, the patients with progressive NBD revealed significant elevation of CSF interleukin-6 activities in comparison with patients with active BD without progressive course. Serum interleukin-6 activities of these groups did not reveal a significant difference [11].

In another investigation in the same center, CSF interleukin-6 levels were used as a biomarker of therapeutic response to low-dose methotrexate. After 12 months, CSF interleukin-6 levels were significantly decreased. However, 6 months after discontinuation of methotrexate, all the patients showed significant exacerbation associated with the marked elevation of CSF interleukin-6 [12].

Increased CSF interleukin-6 level has been also shown in patients studied by Hatachi et al. [13], Nakano et al. [14], Sakuta et al. [21], Watanabe [22], and Hamada [23].

Decreased CSF IL-6 levels with clinical improvement were seen in Sakuta et al.'s [21] reports. In Hirohata's study, the sensitivity and specificity of CSF interleukin-6 for the diagnosis of CP NBD versus the recovery phase of acute NBD were 86.7 and 94.7%, respectively [8].

In Akman-Demir's study, in patients with acute parenchymal NBD, CSF, but not serum IL-6, was significantly increased in comparison to other groups. Patients with chronic progressive parenchymal NBD showed increased CSF IL-6 levels as well, albeit less prominent. Patients with increased CSF IL-6 levels were more possible to have increased CSF cell counts and total protein levels, and these three parameters were correlated with long-term 3-year disease outcome. Interestingly, some patients with chronic course revealed increased CSF levels despite otherwise normal CSF studies.

In Hirohata's study, the CSF levels of IL-6 and IL-8 were significantly elevated in the NB patients compared with those measured in the control patients. There were no significant differences in the CSF levels of IL-6 and IL-8 between the patients with acute NB and those with chronic progressive NB. The CSF levels of IL-6 and IL-8 were significantly decreased following clinical improvement in both acute NBD and chronic progressive NBD patients. The CSF levels of IL-6 were significantly correlated with the CSF levels of IL-8 in patients with acute NB, but not with chronic progressive NBD [8].

In Sumita et al.'s study, CSF IL-6 levels were slightly elevated in patients with NBD [24].

Other CSF Studies

In 1992, Wang et al. showed that anticardiolipin antibodies, especially the IgM isotype, were highly elevated in the CSF of patients with NBD in comparison to those with non-inflammatory neurological diseases. Levels of IgM isotype of anticardiolipin antibodies decreased with clinical improvement [20].

In Aoyama et al.'s study, CSF IgA was shown to be increased [25].

In Hamzaoui et al.'s study, a linear correlation was seen between matrix metalloproteinase-9 and CSF-mononuclear cells in neuro-Behçet patients. There was also an increased matrix metalloproteinase-9/tissue inhibitor of metalloproteinase-1 ratio. They concluded that matrix metalloproteinases may be involved in the pathogenesis of neuro-Behçet by promoting local damage. Matrix metalloproteinases seem to be involved in the leukocyte trafficking within the CNS [26].

In Hamzaoui et al.'s study, the mean CSF vascular endothelial growth factor (VEGF) was significantly higher in neuro-Behçet patients compared to those with non-inflammatory neurological diseases. There was an association between CSF VEGF and leukocyte count in neuro-Behçet patients. A positive correlation was also seen between neuro-CSF vascular endothelial growth factor (VEGF) and CSF $CD4^+$ cells. Vascular endothelial growth factor mRNA was also significantly increased in patients with neuro-Behçet disease compared to patients with non-inflammatory neurological diseases. Increased vascular endothelial growth factor may demonstrate the disturbance of the blood-brain barrier [27].

In Tasci et al.'s study, 65-kD mycobacterial heat shock protein (hsp65), anti-m-hsp65 IgG, and IgM and IgA antibodies in the serum and cerebrospinal fluid of 25 Behçet patients with cerebral parenchymal involvement, seven Behçet patients with intracranial hypertension, eight Behçet patients without central nervous system involvement, 30 patients with multiple sclerosis, and 24 patients with non-inflammatory CNS disorders were measured. CSF anti-hsp65 antibodies were positive in 48% of the patients with parenchymal NBD. This was significantly higher than the positivity rate found in Behçet patients with intracranial hypertension, Behçet patients without CNS involvement, MS patients, and patients with non-inflammatory CNS disorders. CSF IgG and IgM antibodies were high in parenchymal NBD. Meanwhile, serum IgM and IgA responses were increased in BD and intracranial hypertension. They conclude that pathogenesis of parenchymal NBD and non-parenchymal NBD is different [28].

In Hamzaoui et al.'s study, IL-15 levels were increased both in active BD with and without neurological manifestations, which implies that this cytokine might not be very important for CNS disease in Behçet. CSF IL-15 levels in neuro-BD patients and inflammatory neurological disease patients were higher than those of patients with non-inflammatory neurological disease. It is assumed that an elevated CSF/serum IL-15 ratio reflects a local IL-15 production by CNS-infiltrating inflammatory cells. Interleukin-15 is a pro-inflammatory cytokine, produced by activated blood monocytes, macrophages, dendritic cells, and activated glial cells. IL-15 promotes the T-cell proliferation and induction of cytolytic effector cells including natural killer and cytotoxic cells and stimulates B cell to proliferate and secrete the immunoglobulins. IL-15 influences the cellular adhesion and transendothelial inflammation-directed migration of activated T lymphocytes [29].

In Vural et al.'s study, high-titer stress-induced phosphoprotein 1 (STIP-1) antibodies were more frequently detected in parenchymal NBD [30].

In Kawai et al.'s study, CSF β2 microglobulin, serum β2 microglobulin, Q albumin, and CSF β2 microglobulin index were significantly increased in patients with NBD in comparison to the control patients. There were no significant differences in these indices between 11 patients with chronic NBD and six patients with acute NBD. In nine patients with NBD, CSF β2 microglobulin and Q albumin were significantly decreased when the CNS manifestations were improved by successful treatment, whereas CSF β2 microglobulin index and serum β2 microglobulin were not significantly changed. The results indicate that the elevation of CSF β2 microglobulin, which results from the transudation of serum β2 microglobulin as well as the increased intrathecal synthesis presumably by infiltrating lymphocytes, is a good marker of CNS disease activity of NBD. The data, however, also suggest that the intrathecal synthesis of β2 microglobulin might not be parallel to the CNS disease activity of NBD [31].

In Hamzaoui et al.'s study, natural killer T (NKT) cells were increased in the CSF of the NBD patients compared with patients with non-inflammatory neurological disorders. Meanwhile, the reactivity of natural killer T cells in the CSF of the NBD patients was not impaired. CSF natural killer T cells of the NBD patients revealed an increased expression of interferon-γ producing cells, indicating a Th1 cell immunogenesis of NBD [32].

In Hamzaoui et al.'s study, expression of T-box transcription factor 21 (TBX21), RAR related orphan receptor C (RORC), and forkhead box P3 (FOXP3) was increased in NBD patients compared to BD patients presenting only with headache and patients with non-inflammatory neurological disorders. TBX21, RORC, and FOXP3 are related to Th1, Th17, and Treg immune processes, respectively. Epstein-Barr virus-induced gene 3 (EBI3), also known as interleukin-27 subunit beta, and T-helper 2 associated GATA binding protein 3 (GATA3) expressions were found to be decreased in NBD patients. Analysis of transcription factor ratios revealed an increase in the RORC/FOXP3 and TBX21/GATA3 ratios in the NBD patients. They concluded that both Th1 and Th17 mRNA expressions may have a role in deregulation of Treg cells and activation of pro-inflammatory T cell subpopulations [33].

Hamzaoui et al. investigated B cell-activating factor of the tumor necrosis factor family (BAFF) and BAFF-R (BAFF receptor) in NBD, multiple sclerosis, and non-inflammatory neurological diseases (NIND). Cerebrospinal fluid level of BAFF messenger RNA and BAFF-R mRNA in unfractionated cells was measured. BAFF and BAFF-R expression in CSF was increased in NBD and MS patients compared to NIND patients. RNA levels of BAFF and BAFF-R were significantly correlated in NBD and MS patients. Serum soluble BAFF levels were increased in NBD and MS patients, but they did not correlate with BAFF expression in CSF [34].

In Sumita et al.'s study, BAFF levels in patients with NBD were significantly elevated compared with healthy controls, but no statistically significant elevation was shown compared with patients with epidemic aseptic meningitis and multiple sclerosis (MS). CSF BAFF levels did not correlate with the serum BAFF levels, CSF cell counts, or CSF IL-6 levels in patients with NBD. Meanwhile, BAFF levels were significantly increased in those with a slowly progressive course in comparison to patients with acute course. These data suggested that BAFF was produced within the central nervous system and may be linked with pathogenesis of NBD, particularly with a progressive course [35].

In Kaabachi et al.'s study, interleukin-26 was highly expressed in CSF from BD patients in comparison to healthy controls. CSF IL-26 levels revealed linear correlations with IL-17 level and a negative correlation with IL-37. IL-26-stimulated CD4+ T cells and monocytes promote the generation of Th17 (IL-17A, IL-23) and suppress the Treg (IL-10, TGF-β) cytokines [36].

In Saruhan-Direskeneli et al.'s study, the levels of CSF C-X-C motif chemokine 10 (CXCL10), also known as interferon gamma-induced protein 10, were significantly higher in NBD and infectious and/or inflammatory neurological diseases than non-inflammatory neurological diseases (NIND) and MS. Chemokine (C-X-C motif) ligand 8 (CXCL8) or interleukin 8 was increased in NBD compared to NIND. They conclude that NBD has a mediator pattern similar to neuro-infections compared to autoimmune disorders, such as MS. Accordingly, they speculated that currently undetermined infection might be the initiator of the vasculitic process in NBD [37].

Hamzoui et al.'s study showed the expression of proto-oncogene B cell lymphoma 2 (Bcl-2) in the peripheral blood and CSF lymphocytes of BD patients and healthy controls. In BD patients, a significant percentage of T cells showed enhanced expression of Bcl-2, both in peripheral blood and CSF. Mononuclear cells of patients with BD showed increased amounts of Bcl-2 messenger RNA. The occurrence of circulating T lymphocytes with abnormally high Bcl-2 expression may be explained in part by increased in vivo activation levels [38].

In Hamzoui et al.'s study, IL-33 was elevated in the CSF from NBD patients in comparison to the controls. At the mRNA level, IL-33 was also highly expressed. Likewise, nuclear factor kB (NF-kB), which mediates IL-33 transcription, was also increased in comparison to disease controls. Chemokine mRNA (IP-10 and MCP-1) was highly expressed in NBD compared with the disease controls [39].

Conflict of Interest The author has no conflict of interest.

References

1. Borhani-Haghighi A, Pourmand R, Nikseresht A-R. Neuro-Behcet disease: a review. Neurologist. 2005;11(2):80–9.
2. Borhani-Haghighi A, Samangooie S, Ashjazadeh N, Nikseresht A, Shariat A, Yousefipour G, et al. Neurological manifestations of Behcet's disease. Saudi Med J. 2006;27(10):1542–6.
3. Borhani-Haghighi A, Sharifzad HR, Matin S, Rezaee S. The pathological presentations of neuro-Behçet disease: a case report and review of the literature. Neurologist. 2007;13(4):209–14.
4. Nakamura S, Takase S, Itahara K. Cytological examination of cerebrospinal fluid in eight patients with neuro-Behçet's disease. Tohoku J Exp Med. 1980;132(4):421–30.
5. Al-Araji A, Sharquie K, Al-Rawi Z. Prevalence and patterns of neurological involvement in Behçet's disease: a prospective study from Iraq. J Neurol Neurosurg Psychiatry. 2003;74(5):608–13.
6. Domingos J, Ferrão C, Ramalho J, Rodrigues T, Moreira B, Santos E, et al. Characteristics of Neuro-Behçet's disease in a case-series from a single centre in Northern Portugal. Eur Neurol. 2015;73(5–6):321–8.
7. Yoshimura J, Toyama M, Sekihara Y, Tamatani S, Nagai H, Fujita S, et al. Neuro-Behcet disease mimicking a thalamic tumor. No shinkei geka. Neurol Surg. 2001;29(6):527–31.
8. Hirohata S, Kikuchi H, Sawada T, Nagafuchi H, Kuwana M, Takeno M, et al. Clinical characteristics of neuro-Behcet's disease in Japan: a multicenter retrospective analysis. Mod Rheumatol. 2012;22(3):405–13.
9. Hirohata S, Kikuchi H. Changes in biomarkers focused on differences in disease course or treatment in patients with neuro-Behçet's disease. Intern Med. 2012;51(24):3359–65.
10. Miller WW, Konstas D, Gandhy C, Robertson D. Unusual late onset of parenchymal neuro-Behçet disease. Case Rep Neurol Med. 2016;2016:5720319.
11. Hirohata S, Takeuchi A, Miyamoto T. Association of cerebrospinal fluid IgM index with central nervous system involvement in Behçet's disease. Arthritis Rheum. 1986;29(6):793–6.

12. Sharief M, Hentges R, Thomas E. Significance of CSF immunoglobulins in monitoring neurologic disease activity in Behçet's disease. Neurology. 1991;41(9):1398.
13. Gille M, Sindic C, Laterre P, De PH, Hotermans J, Selak I, et al. Neurological involvement as manifestation of Behçet's disease. 4 clinical case reports. Acta Neurol Belg. 1990;90(5):233–47.
14. Jongen P, Daelmans H, Bruneel B, Den Hartog M. Humoral and cellular immunologic study of cerebrospinal fluid in a patient with Behçet encephalitis. Arch Neurol. 1992;49(10):1075–8.
15. McLean B, Miller D, Thompson E. Oligoclonal banding of IgG in CSF, blood-brain barrier function, and MRI findings in patients with sarcoidosis, systemic lupus erythematosus, and Behçet's disease involving the nervous system. J Neurol Neurosurg Psychiatry. 1995;58(5):548–54.
16. Daoudi S, Lounis M, Ait-Kaci-Ahmed M. Parenchymal neuro-Behcet's disease. Clinical and paraclinical characteristics, a report of 40 cases. Presse Med. 2014;43(5):e119–25.
17. Saruhan-Direskeneli G, Yentur SP, Mutlu M, Shugaiv E, Yesilot N, Kurtuncu M, et al. Intrathecal oligoclonal IgG bands are infrequently found in neuro-Behcet's disease. Clin Exp Rheumatol. 2013;31(3 Suppl 77):25–7.
18. Miller WW, Konstas D, Gandhy C, Robertson D. Unusual late onset of parenchymal neuro-Behcet disease. Case Rep Neurol Med. 2016;2016:5720319.
19. Liu HM, Dong C, Zhang YZ, Tian YY, Chen HX, Zhang S, et al. Clinical and imaging features of spinal cord type of neuro Behcet disease: a case report and systematic review. Medicine. 2017;96(40):e7958.
20. Wang CR, Chuang CY, Chen CY. Anticardiolipin antibodies and interleukin-6 in cerebrospinal fluid and blood of Chinese patients with neuro-Behcet's syndrome. Clin Exp Rheumatol. 1992;10(6):599–602.
21. Sakuta K, Sengoku R, Morita M, Matsusima S, Mochio S, Iguchi Y. A case of neuro-Behcet disease presenting with a longitudinal spinal cord lesion without pain. Rinsho shinkeigaku. Clin Neurol. 2014;54(1):16–21.
22. Watanabe M, Kobayashi R, Hasegawa T, Yokoi S, Okada H, Okuda S. A case of chronic progressive neuro-Behcet's disease with cerebellar ataxia and bulbar palsy preceding mucocutaneo-ocular symptoms. Rinsho shinkeigaku. Clin Neurol. 2018;58(2):105–10.
23. Hamada K, Takei R, Sakiyama Y, Moriyama H, Hashiguchi A, Takashima H. A case of chronic progressive neuro-Behcet disease with extensive cerebral atrophy and elevated CSF IL-6 activity treated with infliximab. Rinsho shinkeigaku. Clin Neurol. 2018;58(1):30–4.
24. Akman-Demir G, Tuzun E, Icoz S, Yesilot N, Yentur SP, Kurtuncu M, et al. Interleukin-6 in neuro-Behcet's disease: association with disease subsets and long-term outcome. Cytokine. 2008;44(3):373–6.
25. Aoyama J, Inaba G, Shimizu T. Third complement component in cerebrospinal fluid in neuro-Behcet's syndrome. Conversion patterns by crossed immunoelectrophoresis. J Neurol Sci. 1979;41(2):183–90.
26. Hamzaoui K, Maitre B, Hamzaoui A. Elevated levels of MMP-9 and TIMP-1 in the cerebrospinal fluid of neuro-Behcet's disease. Clin Exp Rheumatol. 2009;27(2 Suppl 53):S52–7.
27. Hamzaoui K, Ayed K, Hamza M, Hamzaoui A. VEGF and mRNA VEGF expression in CSF from Behcet's disease with neurological involvement. J Neuroimmunol. 2009;213(1–2):148–53.
28. Tasci B, Direskeneli H, Serdaroglu P, Akman-Demir G, Eraksoy M, Saruhan-Direskeneli G. Humoral immune response to mycobacterial heat shock protein (hsp)65 in the cerebrospinal fluid of neuro-Behcet patients. Clin Exp Immunol. 1998;113(1):100–4.
29. Hamzaoui K, Hamzaoui A, Ghorbel I, Khanfir M, Houman H. Levels of IL-15 in serum and cerebrospinal fluid of patients with Behcet's disease. Scand J Immunol. 2006;64(6):655–60.
30. Vural B, Ugurel E, Tuzun E, Kurtuncu M, Zuliani L, Cavus F, et al. Anti-neuronal and stress-induced-phosphoprotein 1 antibodies in neuro-Behcet's disease. J Neuroimmunol. 2011;239(1–2):91–7.
31. Kawai M, Hirohata S. Cerebrospinal fluid beta(2)-microglobulin in neuro-Behcet's syndrome. J Neurol Sci. 2000;179(S 1–2):132–9.

32. Hamzaoui K, Kamoun M, Houman H, Hentati F, Hamza M, Ayed K, et al. Discrepancies of NKT cells expression in peripheral blood and in cerebrospinal fluid from Behcet's disease. J Neuroimmunol. 2006;175(1–2):160–8.
33. Hamzaoui K, Borhani Haghighi A, Ghorbel IB, Houman H. RORC and Foxp3 axis in cerebrospinal fluid of patients with neuro-Behcet's disease. J Neuroimmunol. 2011;233(1–2):249–53.
34. Hamzaoui K, Houman H, Hentati F, Hamzaoui A. BAFF is up-regulated in central nervous system of neuro-Behcet's disease. J Neuroimmunol. 2008;200(1–2):111–4.
35. Sumita Y, Murakawa Y, Sugiura T, Wada Y, Nagai A, Yamaguchi S. Elevated BAFF levels in the cerebrospinal fluid of patients with neuro-Behcet's disease: BAFF is correlated with progressive dementia and psychosis. Scand J Immunol. 2012;75(6):633–40.
36. Kaabachi W, Bouali E, Berraies A, Dhifallh IB, Hamdi B, Hamzaoui K, et al. Interleukin-26 is overexpressed in Behcet's disease and enhances Th17 related -cytokines. Immunol Lett. 2017;190:177–84.
37. Saruhan-Direskeneli G, Yentur SP, Akman-Demir G, Isik N, Serdaroglu P. Cytokines and chemokines in neuro-Behcet's disease compared to multiple sclerosis and other neurological diseases. J Neuroimmunol. 2003;145(1–2):127–34.
38. Hamzaoui K, Hamzaoui A, Zakraoui L, Chabbou A. Expression of Bcl-2 in inflammatory sites from patients with active Behcet's disease. Mediat Inflamm. 1999;8(2):101–6.
39. Hamzaoui K, Borhani-Haghighi A, Kaabachi W, Hamzaoui A. Increased interleukin 33 in patients with neuro-Behcet's disease: correlation with MCP-1 and IP-10 chemokines. Cell Mol Immunol. 2014;11(6):613–6.

Imaging of Neuro-Behçet's Disease

Mehmet Barburoğlu, Pınar Beyaz, and N. Serra Sencer

Introduction

Neuro-Behçet's disease (NBD) is usually defined as the occurrence of neurologic symptoms in a patient who has been diagnosed with *Behçet's disease (BD),* which cannot be explained by other systemic or neurologic disease. The prevalence has been reported in the literature with a wide range going as high as 50%; however, most larger series report the prevalence to be between 3% and 9% [1–3]. Longitudinal follow-up of large cohorts over time shows increase in the prevalence of central nervous system (CNS) involvement in BD [4]. Neuroimaging is a crucial part of the diagnosis in NBD. Neuroimaging diagnosis is usually not very complicated since imaging findings are well defined and easy to recognize and correlate with the symptoms in a patient with known BD. A neurologic onset in BD is not likely, and even in cases where the first symptoms seem to be related to the central nervous system, a thorough history and examination often reveals oral ulcers or other systemic manifestations of BD dating back to months to years before the neurological symptoms [5]. The presence of abnormal imaging findings suggestive of NBD (imaging evidence) has been suggested as one of the diagnostic criteria for NBD, in addition to fulfilling the International Diagnostic Criteria for BD and onset of neurological symptoms not otherwise explained by systemic or neurologic disease [6]. Brain stem and/or corticospinal tract syndromes or signs of raised intracranial pressure in a patient with BD often call for a cranial magnetic resonance imaging study with contrast medium and magnetic resonance venography (MRV) to exclude parenchymal or vascular involvement

M. Barburoğlu (✉) · N. S. Sencer
Istanbul University, Istanbul Faculty of Medicine, Department of Radiology, Division of Neuroradiology, Istanbul, Turkey

P. Beyaz
Klinikum Bremen Mitte, Interventional and Diagnostic Neuroradiology Department, Bremen, Germany

E. Tüzün, M. Kürtüncü (eds.), *Neuro-Behçet's Disease*,
https://doi.org/10.1007/978-3-030-55273-2_10

of the central nervous system. These studies will often provide imaging diagnosis, although further studies may rarely be necessary.

Neuroimaging Tools in NBD

Historically, *computed tomography (CT)* has been the first diagnostic imaging method used in NBD, and CT findings have been reported in a few large patient series [7, 8]. Despite advances in CT technology, its sensitivity remains inferior to MRI for parenchymal NBD involvement. CT angiography (CTA), on the other hand, emerges as a fast, precise, and informative technique in the evaluation of vascular involvement of NBD. CT angiography can be performed rapidly using iodinated contrast medium applied in a bolus fashion intravenously, and excellent axial source images, multiplanar reconstructions, and three-dimensional volume-rendered images can be generated from the desired region of arterial or venous structures. Although MR imaging can provide parenchymal and vascular images at the same setting at high resolution without ionizing radiation or use of contrast medium in NBD, CTA can still be used in cases where MR angiography or MRV is not diagnostic or in patients with claustrophobia or critical illness where a rapid vascular diagnosis is necessary [9]. CTA is often considered a more precise technique for preoperative or pre-endovascular treatment evaluation of Behçet's disease-related aneurysms because of better delineation of vascular structures and relationship to bony and skull base structures. CTA can also be performed via flat detector angiography units at the setting of catheter angiography for detailed investigation of brain stem or white matter venules which are significant regions of histopathological NBD involvement [10].

MRI of the brain with gadolinium contrast agent is the diagnostic imaging method of choice in NBD. MR venography (MRV) of the brain (and neck) as well as MR angiography (MRA) of the brain (and neck) may be indicated in patients in whom there is clinical suspicion for vascular (venous or arterial) involvement. MRI offers the highest sensitivity and specificity in the morphologic detection of lesions of the neural tissue in NBD and has the capability for functional and biochemical analysis with advanced techniques. A standard brain imaging protocol with T1, T2, FLAIR (fluid-attenuated inversion recovery) series, and SWI (susceptibility-weighted imaging) or gradient echo (GRE) sequences (both sensitive to blood products) often completed with T1 W contrast-enhanced T1 series is often sufficient for diagnosis of parenchymal lesions. These sequences can readily identify areas of edema, hemorrhage, and contrast enhancement and show subsequent mass effect or areas of residual encephalomalacia in chronic cases. More recently, DWI has been added to the imaging protocol at most centers. DWI is helpful in detecting areas of restricted diffusion due to acute ischemia and distinguishing between types of edema in corroboration with imaging sequences. Magnetic resonance spectroscopy (MRS) can also be used where differential diagnosis from other diseases (tumor, infection, or inflammation) warrants advanced imaging. MRA and MRV are usually performed using time-of-flight (TOF) technique, and contrast injection with gadolinium contrast agents is used in some centers to enhance the sensitivity of MRV and in the acquisition of MR angiography of large neck arteries. Cerebrospinal fluid (CSF) flow investigations using cardiac triggered phase contrast (PC) MR sequences have also been reported in NBD patients [11].

In NBD cases with spine involvement, T2, T1, and STIR (short tau inversion recovery, for fat-suppressed heavily T2-weighted images) sagittal and axial series followed by gadolinium enhanced T1 sequences in the axial and sagittal planes provide greatest benefit in detecting and characterizing spinal cord lesions. This imaging protocol is often sufficient in the delineation of areas of edema and contrast enhancement in the spinal cord as well as cauda equina fibers.

SPECT (single-photon emission computerized tomography) and PET (positron emission tomography) studies may be utilized in NBD to show changes in regional cerebral perfusion as well as blood flow and oxygen consumption in brain stem and hemispheric lesions [5, 12].

Catheter angiography is the gold standard of vascular imaging. Therefore, any vascular problem not optimally solved with noninvasive vascular imaging calls for cerebral (or spinal) digital subtraction angiography (DSA). This technique will not only provide unequivocal information about the nature of vascular involvement, especially arterial aneurysms, dissections, and arterial or venous occlusions related to NBD, but also enable emergent or elective endovascular treatment of some of these vascular problems [13]. Since the diagnostic yield of DSA may be lower in the elucidation of vasculitic changes in the smaller arteries or venules, it is usually reserved for cases where noninvasive vascular imaging has proved inadequate or when there is need for endovascular or surgical intervention. Some authors have also used DSA for the monitorization of treatment effects [14]. Complications of catheter angiography such as rare groin puncture site problems and one case of rebleeding during arteriography have been reported; nevertheless, in the authors' opinion, catheter angiography is a safe and invaluable procedure for the diagnosis and endovascular treatment of BD-related vascular problems in the hands of experienced interventional neuroradiologists [15].

Histopathology of NBD, Patterns of Central Nervous System Involvement, and the Imaging Counterparts

The pathogenesis of BD is globally considered to be an autoimmune vasculitis predisposed by genetic determinants and triggered by exogenous factors [16]. NBD shares the same pathological features and has two main forms of involvement in the central nervous system:

- *Parenchymal involvement (PNBD)* which favors brain stem as well as cerebral/cerebellar hemispheres, spinal cord, and meninges. The most common specific sites of PNBD are mesodiencephalic junction, cerebellar hemispheres, and basal

ganglia. Periventricular, spinal cord, subcortical cerebral, and optic nerve involvement have also been reported.
- *Non-parenchymal or vascular involvement (NPBD)* which includes dural sinuses (thrombosis) and arteries (occlusion and aneurysm formation) [17].

The main pathological feature common to many studies in NBD is perivascular infiltration of mononuclear, polymorphonuclear, and rarely eosinophilic cells. Areas of focal softening with loss of neurons and myelin are present especially in the brain stem. The earliest report came from Berlin in 1944, stating multiple foci of cellular infiltration within the meninges and brain parenchyma [18]. Following reports, autopsy and some biopsy results were mostly in agreement with these findings; however, presence of necrosis, demyelination, and even perivascular cuffing were not common to all studies. Some rare case reports described isolated demyelination or gliosis without perivascular cellular infiltration [19–24]. Perivascular inflammation was perivenular, periarterial, and perivenular or periarterial with varying intensities. The prominent involvement of the brain stem was unanimously attributed to the lack of collateral venous system in this anatomical region, facilitating local venous obliteration and ischemia or even hemorrhage. There is clearly a histopathological involvement pattern with commonly encountered findings that may prove very useful in complicated cases where pathological diagnosis is mandatory due to insufficient or confusing clinical and imaging findings. Yamada and colleagues have recently reported a patient who presented with neurological involvement and was diagnosed with BD and NBD via brain biopsy of the right frontal tumor-like lesion [25]. On the other hand, diagnosis of NBD still depends on history, physical examination, imaging, and CSF studies, probably due to scarcity of potentially high-risk brain stem biopsies.

Pathology often shows small vessel thrombosis at areas of perivascular infiltration, and the same infliction may result in vasa vasorum vasculitis and concomitant dural sinus thrombosis [25].

Imaging counterpart of Behçet vasculitis is usually T2, FLAIR, and DWI and ADC hyperintensity which is mostly compatible with vasogenic edema. DWI can distinguish cytotoxic and vasogenic edema and help differentiate between acute infarction and inflammation which may be relevant in NBD. Neuronal "loss" has also been implied in a small number of MRS studies by the decreased peak of N-acetylaspartate (NAA) which is considered a marker for neuronal integrity. Blood-brain barrier is usually disrupted in the involved region, resulting in contrast enhancement. The lesions usually resolve with minimal residue. The residue may be infarctions which may follow intense perivascular infiltration disrupting flow or astroglial proliferation [26]. Atrophy, which is a relatively common finding in pathological studies, is easily appreciated on imaging in chronic cases. The meningeal cellular infiltration especially in cases with acute meningeal syndrome presents on imaging as enhancement in the leptomeninges [27]. In chronic cases, meningeal cellular infiltration may change into meningeal fibrosis. Both meningeal cellular infiltration and fibrosis may not have significant imaging counterparts.

Hemorrhage observed in imaging correlates to "diapedesis of red blood cells around veins" and not to arterial rupture according to histopathological investigations. Hemorrhage is detected as areas of low signal on GRE and SWI series and has variable signal intensities on T2, T1, and FLAIR sequences depending in the age of blood products. In NBD, hemorrhage often presents as focal areas of blood products in large edematous regions and rarely as large hematomas; however, frank hemorrhagic lesions as initial radiological presentation of NBD have also been reported [9, 21].

Neuro-Behçet's Disease in Children

BD rarely affects children and adolescents. It usually presents at the beginning of the second decade of life. NBD occurs in 20–25% percent of children with BD and is related to poor prognosis [28]. Symptoms in children with NBD are similar to adults. In a small series, 40% of children with NBD had headache, and another 20% had intracranial hypertension [29]. Brain stem involvement in the mesodiencephalic junction is the prevalent form in children similar to adults. Edematous lesions, areas of punctate and patchy enhancement in the parenchymal and spinal lesions, have also been reported in children [30]. The non-parenchymal form of NBD, which presents mostly as dural venous sinus occlusion, tends to involve younger BD patients, some of whom are children. This form also has a more favorable disease progress and prognosis. Uluduz and colleagues have reported dural sinus thrombosis as the most common form of involvement in pediatric NBD [31]. Children with BD may be more difficult to diagnose; therefore, a child (especially coming from a country located on the Silk Route) diagnosed with intracranial cerebral venous thrombosis should be investigated for BD and NBD.

Patterns of Involvement in NBD and Neuroradiological Findings

There are a few larger series of NBD patients reporting CT findings; however, almost all contemporary imaging literature is based on MR findings [7, 8].

Two patterns of involvement in the CNS are *parenchymal* and *non-parenchymal or vascular*. While parenchymal lesions are located in the brain and spinal cord and rarely in the peripheral nervous system, non-parenchymal lesions mostly involve dural sinuses and infrequently arteries.

Parenchymal lesions mostly present with neurological findings functionally corresponding to the area of involvement, whereas dural sinus occlusions manifest with signs of raised intracranial pressure. The former should be evaluated with contrast-enhanced multi-sequential MR investigation and the latter with an additional MRV study. When there is arterial involvement, presentation depends on the sequelae of the arterial injury (occlusion, dissection, aneurysm formation) and may be in the form of arterial infarction or intracerebral or subarachnoid hemorrhage. In the acutely ill patient with suspicion of intracranial bleeding, the initial exam at most centers is a non-enhanced brain CT scan. Depending on the neurological state of the patient, the degree of emergency, and the treating institution, noninvasive vascular imaging (MRA or CTA) or DSA is immediately performed following initial scan showing intracranial bleeding.

Parenchymal Involvement in Neuro-Behçet's Disease

The most frequent region of involvement in NBD is the brain stem (midbrain, pons, and bulbus), followed by thalami and basal ganglia-internal capsules, white matter, and cerebellum. Telencephalon, optic nerves, and the spinal cord (cervical more than dorsal) are more rare sites of parenchymal disease. Lesions are usually multiple and multicentric; however, single and atypical lesions are also possible.

Acute and subacute lesions are markedly hyperintense on T2 and FLAIR and iso−/hypointense on T1 with hazy borders and become slightly T2 hyperintense and T1 hypointense with good demarcations in chronic cases.

Most larger clinical series have reported a unilateral (or more rarely bilateral) *mesodiencephalic (MDJ)* lesion, which is a T2 and FLAIR hyperintense signal extending from the midbrain upward into the thalami as well as internal capsule and basal ganglia and downward toward the ponto-bulbar area, as the most common intracranial lesion (Fig. 1). The downward extension is along the pontine tegmentum and cerebellar peduncles or along the corticospinal tract in more anterior ones. Sparing of the red nucleus was reported as a common feature in downward extensions [5, 9, 12]. Involvement of the corticospinal tract in the brain stem and internal capsule correlates well with the pyramidal signs commonly encountered in these patients [32]. Erdem and colleagues have noted an enhancing lesion in the internal capsule as the reason for hemiparesis in their patient [33].

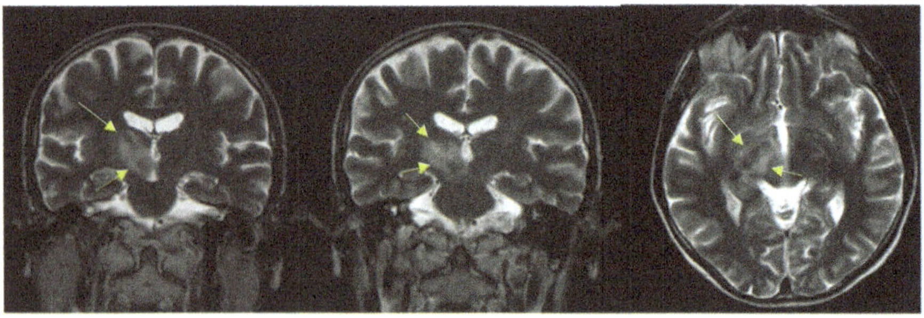

Fig. 1 A right mesodiencephalic junction lesion; note that the lesion originates from the mesencephalon and extends upward into the thalamus, internal capsule, and basal ganglia

Acute or subacute lesions present with the largest diameter of involvement, hazy borders, occasional hemorrhagic signal, and contrast enhancement, which tend to resolve with small, well-demarcated T1 hypointense, T2 slightly hyperintense residue. The acute lesion is attributed to vasogenic (interstitial) edema that is common in venous problems, also characterized by its reversibility and morphologic features on MRI. Hemorrhage is mostly encountered in parenchymal lesions of the brain stem and attributed to red cell diapedesis from venules in the affected area which is best delineated on SWI sequence that can also show venous collaterals. Brain stem, being devoid of collateral longitudinal and transverse collateral venules, is severely affected by the venular attacks of Behçet pathology and is the most common area showing edematous and hemorrhagic changes in this disease (Fig. 2). Hemorrhage signal is permanent on MRI especially on blood product sensitive series such as GRE and SWI. Contrast enhancement is due to breakdown of blood-brain barrier and mostly a radiologic event of the acute and subacute stage and not present in chronic lesions and following treatment. Atrophy, on the other hand, is a frequent radiological finding in the chronic cases, especially in the brain stem. Several studies have reported brain stem atrophy without or more enhanced than cerebral cortical atrophy as a distinguishing feature of NBD [32, 34]. Ventricular enlargement has also been reported on MR scans in chronic NBD [35].

The radiological progression of disease from the acute into the chronic stage has been well depicted, and the tendency to resolve and decrease in size of the lesions over time has been reported by many series (Fig. 3). This has been flanked by MR spectroscopy findings in a few series, which showed decrease in the N-acetylaspartate (NAA) peak at 2 ppm and increase in the creatine (Cr) and choline (Cho) peaks during the acute stage [36, 37]. NAA is a marker for neuronal and axonal integrity, and NAA peak decrease has been reported to be reversible by Nussel and colleagues after treatment. This may be considered, albeit by very few studies, as biochemical data suggesting lesion resolution after therapy or simply remission. This "reversibility" of the lesions is considered a distinguishing feature of NBD useful in the differential diagnosis from other inflammatory and tumoral conditions of the CNS.

"Silent lesions" have also been implicated in NBD. In a recent prospective series of progressive or relapsing-remitting NBD cases with a mean follow-up of 29 months, increase in the number of MRI lesions and evolution of black holes have

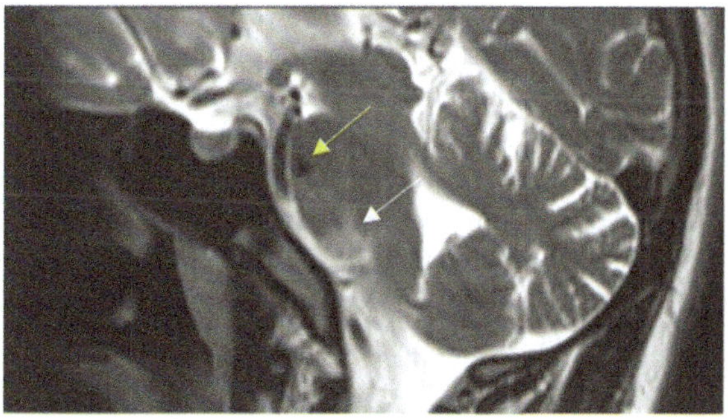

Fig. 2 T2 sagittal image in a 32-year-old male patient with NBD shows a hypointense hemorrhagic lesion in the ventral portion of pons (yellow arrow). The patient also has an edematous lesion in the ponto-medullary junction (white arrow)

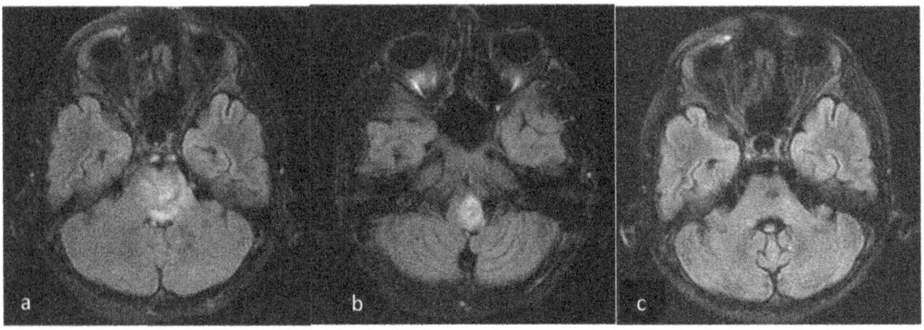

Fig. 3 Reversibility of lesions in NBD; a 43-year-old male patient with known BD; the image shows a confluent central lesion in the brain stem (**a**, **b**). Repeat MR scan at 8 months (**c**) shows near total resolution of the lesion except for a small residue

been noted [38]. A perfusion MR study has also shown that alongside conventional MR lesion sites, other regions in the brain show hypoperfusion, suggesting that subclinical vascular and neuronal dysfunction may be more widespread than clinically expected [39].

A "tumor-like" presentation of parenchymal NBD has been reported [40–46]. These are usually case reports which have been misdiagnosed as brain tumors because of misleading or nonexistent clinical diagnosis of BD/NBD and the imaging appearance suggestion of tumor. Biopsy, open surgery, or advanced MR imaging (MR spectroscopy, diffusion tensor imaging (DTI)) in similar cases with "tumor-like presentation" usually provides precise diagnosis; however, the key to avoiding unnecessary surgical intervention with such patients is to evaluate the patient's clinical and laboratory findings very carefully to exclude NBD and other non-tumoral systemic diseases which may involve the CNS and mimic tumor. These patients with isolated large lesions of the brain parenchyma have been shown to have worse prognosis and more long-term disability due to disease than patients with more "classical" form of NBD [47].

Recurrent meningitis or meningeal syndrome has been reported as one of the rare NBD presentations [48, 49]. Although more frequent findings of NBD coexistent in these patients facilitate diagnosis, isolated and/or recurrent aseptic meningitis should incite investigation for BD/NBD in cases that have not been diagnosed previously. Kara and colleagues have reported a child with acute meningeal syndrome and leptomeningeal enhancement on brain MR, who was subsequently diagnosed with BD according to International Study Group of Behçet's Disease criteria for diagnosis. Although leptomeningeal enhancement characteristic for meningitis is very rare, as stated before, microscopic cellular infiltration of the meninges is quite frequent in histopathological studies. Two studies have reported thickening and enhancement in the tentorium and falx in their patients with NBD; therefore, "pachymeningitis" can also be considered as a suggestive radiological finding in NBD [50, 51].

NBD may present with periventricular and spinal lesions which may suggest acute disseminated encephalomyelitis or multiple sclerosis in patients who have not been previously diagnosed [52, 53]. The differential diagnosis of diseases which may mimic NBD will be discussed below. Optic myelopathy which is a rare finding in NBD may also cause misdiagnosis in some cases. Kocer and colleagues have followed one patient with optic myelopathy and NBD with optic nerve enhancement on the afflicted side in the acute phase and neural atrophy on follow-up [9]. NBD may share some clinical and radiological features with neuromyelitis optica which should be included in the differential diagnosis.

White matter lesions in NBD are usually small foci of high T2 signal scattered everywhere in the white matter without predilection for periventricular and callosal septal areas (Fig. 4). They can be extensive and confluent as well. As white matter lesions in NBD are rarely encountered in isolation, other lesions are helpful in differential diagnosis.

Spinal involvement in NBD is not common but has been reported in many series and case reports [9, 52, 53]. Cervical spinal cord is the most common segment to be involved. The lesion often causes enlargement in the involved area in the acute phase and continues more than two vertebral segments or may ascend into the brain stem in a myelitis-like fashion (Fig. 5). In cervical cord lesions, upward extension is usually through sensory long fiber tracts. Much like their intracranial counterparts, cranial and caudal extensions of spinal cord lesions are most pronounced during the acute phase and diminish over time leaving smaller residues. Parenchymal lesions of the spinal cord may be single or multiple. Not all lesions extend along long segments; small and nodular or patchy lesions have also been reported. Nodular or patchy enhancement can be seen in spinal cord lesions in the acute and subacute phases gradually diminishing over time or during treatment. Recurrent transverse myelitis has been encountered in a young male patient treated with infliximab for BD [54].

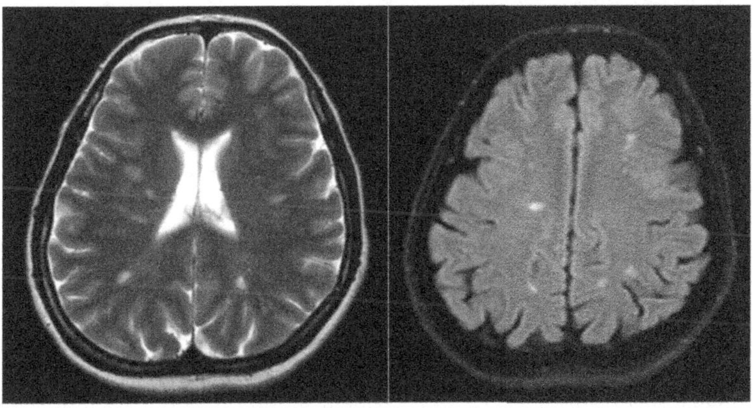

Fig. 4 Multiple white matter lesions in an NBD patient. Note that the lesions are widely distributed with no propensity for periventricular or callosal septal interphase

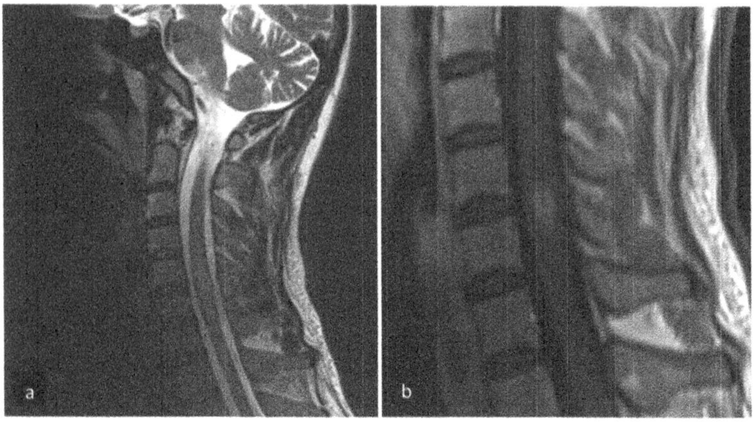

Fig. 5 Spinal involvement in NBD; T2 W sagittal image of the cervical spine shows mild fusiform enlargement of the cord between C4 and C6, caused by a myelitis-like lesion (**a**). Enlarged sagittal T1 W contrast-enhanced study of the same patient reveals homogenous nodular enhancement in the center of the lesion (**b**)

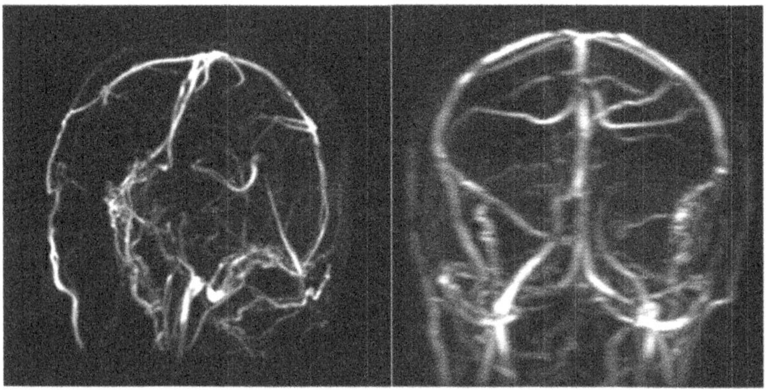

Fig. 6 Extensive sinus thrombosis in an NBD patient; both transverse sinuses are occluded; there are also significant collateral venous channels due to long-standing venous occlusion

Non-parenchymal Involvement in Neuro-Behçet's Disease

Around one-third of BD patients with CNS involvement present with cerebral venous sinus thrombosis [4] (Fig. 6). This type of CNS involvement has a more insidious course dating back to several weeks to months before admission, is more frequent in young male adults with BD, and offers a more favorable prognosis. Relapse, on the other hand, is substantial. Concomitant parenchymal and non-parenchymal involvement in the same patient is rare [1, 55]. The most common sinus to get occluded is superior sagittal sinus, followed by the transverse sinus. One or more sinuses may be involved. When the dural sinus thrombosis is propagated into the cortical veins, venous infarctions and hemorrhage may ensue. Nevertheless, this is quite rare in NBD. Indolent course, together with the relative rarity of venous infarctions, has led to the opinion that cerebral venous thrombosis in NBD may not be acute and complete in most patients [5].

MR or CT venography of the brain and neck veins will be sufficient in the definitive diagnosis of cerebral venous thrombosis in the vast majority of cases. It should be kept in mind that especially in the diagnosis of acute thrombosis, T1, T2, and DWI series are very helpful and should be acquired and consulted in addition to venography images in these patients. MR and MRV studies are not only proficient in diagnosing thrombosis but also very useful in the monitorization of treatment. Neuroimaging follow-up timing and protocol should be tailored according to the patient's characteristics and treatment regimen but is considered safe in this relatively young patient group since it does not entail use of ionizing radiation or contrast medium. DSA is rarely necessary in the event of venous thrombosis, except for assessment of intracranial hemodynamics in advanced cases, or when endovascular treatment is necessary.

Aside from cerebral venous thrombosis, an "arterial form" of NBD has also been defined. There have been observations of internal carotid artery or vertebral artery occlusions, vertebral artery dissections, and intracranial aneurysms and arteritis in BD patients with corresponding neurological sequelae [5] (Fig. 7). Intracranial hemorrhage due to these lesions may be seen; however, hemorrhage is much more frequently secondary to parenchymal lesions.

In treatment of intracranial aneurysms and major artery involvement, a multidisciplinary approach is important. Post-infectious or mycotic aneurysm in a BD patient should be treated by management of the underlying disease; however, an incidental saccular brain aneurysm can be treated at an experienced endovascular or neurosurgical institution in similar fashion to a non-BD patient (Fig. 8). Management

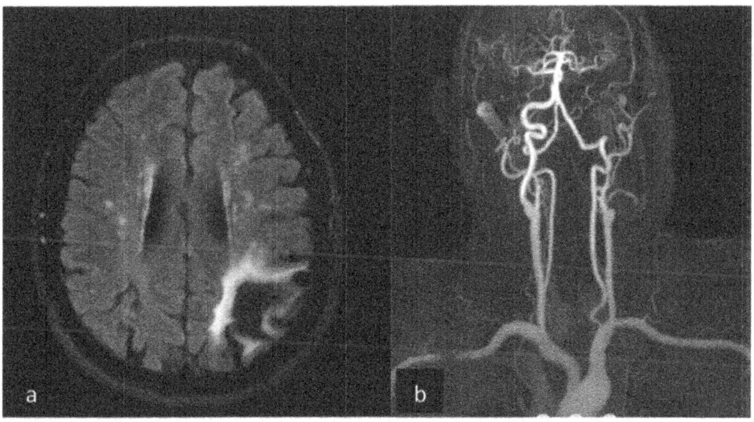

Fig. 7 Ischemic stroke in an NBD patient due to internal carotid artery (ICA) occlusion; the patient has a left parietal infarction as well as multiple white matter lesions due to small vessel disease (**a**); MRA of the neck vessels shows total occlusion of the left ICA (**b**)

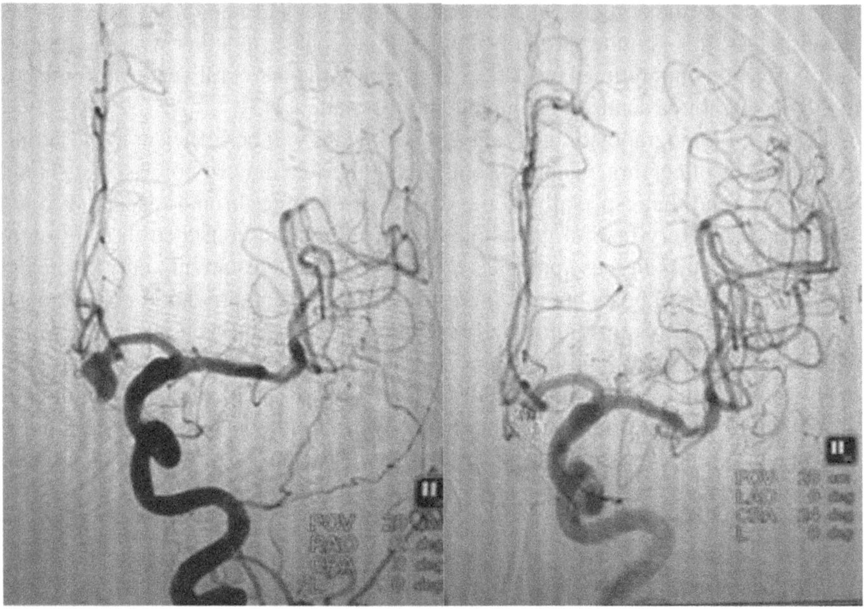

Fig. 8 A 58-year-old male BD patient has been admitted with subarachnoid hemorrhage. DSA image on the left shows a (ruptured) anterior communicating artery aneurysm which was treated by primary coiling

of large vessel involvement in the head and neck is also a controversial issue, and treatment for underlying disease remission is very important, although endovascular treatment with angioplasty and stents or thrombectomy may also be called for. Because of disease characteristics and vulnerabilities of the Behçet patient, endovascular or surgical intervention must be undertaken following deliberation by experienced teams at reference centers.

Radiological Differential Diagnosis

Neurological disease is rarely the initial manifestation of BD; even in cases that seem so, careful history taking and physical examination often reveal hints about the underlying disease. Still, differential diagnosis can be crucial, and neuroradiological findings may be helpful.

A typical patient with NBD is a young individual who often presents with an acute/subacute brain stem syndrome as well as hemiparesis; therefore, prompt differential diagnosis from multiple sclerosis (MS) and stroke may be warranted.

MS is more likely to be seen in young females, with optic neuritis, sensory symptoms, and spinal involvement which are rarer in NBD. Intracranial MS lesions tend to involve periventricular white matter with propensity of callosal septal interphase location as multiple small lesions, while NBD often presents with nonspecific locations of white matter lesions often in combination with brain stem lesions. While brain stem lesions in NBD prefer MDJ or central pons and corticospinal tract, MS tends to involve the floor of the fourth ventricle and middle cerebellar peduncle. Spinal lesions in MS are small and widely scattered in the cord, but NBD lesions involve long segments of the cord in a myelitis-like fashion. On axial images, NBD mostly involves central gray matter, while MS creates eccentric, off midline plaques [5, 12]. In chronic cases, atrophy in NBD tends to be isolated to the brain stem, whereas MS has both cerebral and brain stem atrophy.

An acute stroke-like presentation in a young patient may necessitate a thorough clinical and laboratory investigation for vasculitic disorders which may involve the CNS. The NBD lesions on cross-sectional imaging are not compatible with arterial territories and often present with vasogenic instead of cytotoxic edema, and vascular imaging of the head and neck arteries is often negative. İschemic stroke at the acute phase shows cytotoxic edema compatible with arterial territories, and vascular imaging will often provide significant information about the underlying etiology.

Sarcoidosis is a disease which may also present with uveitis, arthritis, and CNS involvement and must at times be differentiated from NBD. Cranial neuropathies mostly involving the seventh nerve and diabetes insipidus are frequent in sarcoidosis. The granulomatous lesions of sarcoidosis may involve the meninges, the suprasellar cistern, and the infundibular stalk and produce thick and slightly nodular enhancement in these regions. There is also a meningoencephalitic form with parenchymal lesions. These are usually quite distinctive from NBD lesions [5].

Brain stem gliomas and primary central system lymphoma may be included in the differential diagnosis of pontine or MDJ lesions in NBD. Lymphomas tend to restrict diffusion and show more enhancement than NBD vasculitis and show remarkable and fast resolution after steroid treatment. Gliomas have varying diffusion and contrast enhancement properties depending on their histopathological grade; however, conventional MR characteristics in combination with MRS features suggestive of gliomas are often sufficient for diagnosis. Gliomas tend to show enhanced myoinositol peak due to their astrocytic origin and heightened choline peak due to increased cell turnover. Higher-grade tumors exhibit lactate and lipid peaks which are due to anaerobic metabolism and myelin breakdown [32].

Tuberculosis (TB) is a systemic disease like BD and has similar propensity for CNS involvement. The two diseases can easily be differentiated on the basis of clinical and laboratory findings. TB's classical meningoencephalitis produces a thick enhancing exudate in the basal cisterns often causing vascular and cranial nerve complications; the second most common form of CNS disease is tuberculomas, which are often multiple, ring-enhancing nodular lesions with peripheral edema.

Systemic lupus erythematosus (SLE) and primary antiphospholipid syndrome (PAS) may also need to be differentiated from NBD at rare instances, and neuroimaging in these diseases often shows arterial infarctions and gray matter changes which are rare in NBD. SLE and PAS rarely involve the brain stem which is often involved in NBD. The rate of cerebral atrophy is high in SLE due to steroid treatment. NBD causes atrophy in the brain stem more often than cerebral hemispheres.

References

1. Akman-Demir G, Serdaroglu P, Tasçi B. Clinical patterns of neurological involvement in Behçet's disease: evaluation of 200 patients. The Neuro-Behçet Study Group. Brain. 1999;122(Pt 11):2171–82.
2. Kidd D, Steuer A, Denman AM, Rudge P. Neurological complications in Behçet's syndrome. Brain. 1999;122(Pt 11):2183–94.
3. Davatchi F, Shahram F, Chams-Davatchi C, Shams H, Nadji A, Akhlaghi M, Faezi T, Ghodsi Z, Larimi R, Ashofteh F, Abdollahi BS. Behcet's disease in Iran: analysis of 6500 cases. Int J Rheum Dis. 2010;13(4):367–73.
4. Kürtüncü M, Tüzün E, Akman-Demir G. Behçet's disease and nervous system involvement. Curr Treat Options Neurol. 2016;18:19.
5. Saip S, Akman-Demir G, Siva A. Neuro-Behçet syndrome. Handb Clin Neurol. 2014;121:1703–23.
6. Siva A, Saip S. The spectrum of nervous system involvement in Behçet's syndrome and its differential diagnosis. J Neurol. 2009;256(4):513–29.
7. Inaba G. Clinical features of neuro-Behcet syndrome. In: Lehner T, Barnes CG, editors. Recent advances in Behcet's disease. International congress and symposium series number 103. London: Royal Society of Medicine Services; 1986. p. 235–46.
8. Siva A, Necdet V, Yurdakul S, Yardim M, Denktas F, Yazici H. Neuro-radiologic findings in neuro-Behcet syndrome. In: O'Duffyl D, Kokmen E, editors. Behcet's disease: basic and clinical aspects. New York: Dekker; 1991. p. 323–9.
9. Koçer N, Islak C, Siva A, Saip S, Akman C, Kantarci O, Hamuryudan V. CNS involvement in neuro-Behçet syndrome: an MR study. AJNR Am J Neuroradiol. 1999;20(6):1015–24.
10. Alis D, Civcik C, Erol BC, Kizilkilic O, Kocer N, Islak C. Flat-detector CT angiography in the evaluation of neuro-Behçet disease. Diagn Interv Imaging. 2017;98(11):813–5.
11. Ünlü S, Doğan M, Kapicioğlu Y, Kamişli S, Öner S, Yildirim IO, Öztürk M. CSF flow patterns in the brain in patients with neuro-Behçet disease and Behçet disease. Eur Rev Med Pharmacol Sci. 2017;21(17):3906–10.
12. Borhani Haghighi A, Pourmand R, Nikseresht AR. Neuro-Behçet disease. A review. Neurologist. 2005;11(2):80–9.
13. Kizilkilic O, Albayram S, Adaletli I, Ak H, Islak C, Kocer N. Endovascular treatment of Behçet's disease-associated intracranial aneurysms: report of two cases and review of the literature. Neuroradiology. 2003;45(5):328–34.
14. Krespi Y, Akman-Demir G, Poyraz M, Tugcu B, Coban O, Tuncay R, Serdaroglu P, Bahar S. Cerebral vasculitis and ischaemic stroke in Behçet's disease: report of one case and review of the literature. Eur J Neurol. 2001;8(6):719–22.
15. Aktaş EG, Kaplan M, Ozveren MF. Basilar artery aneurysm associated with Behçet's disease: a case report. Turk Neurosurg. 2008;18(1):35–8.
16. Davatchi F, Shahram F, Akbarian M, et al. Behcet disease: analysis of 3443 cases. APLAR J Rheumatol. 1997;1:2–5.
17. Borhani Haghighi A, Sharifzad HR, Matin S, Rezaee S. The pathological presentations of neuro-Behçet disease: a case report and review of the literature. Neurologist. 2007;13(4):209–14.
18. Berlin C. Behcet syndrome with involvement of the central nervous system. Arch Derm Syphilol. 1944;49:227–33.
19. Silfverskiold BP. Recurrent uveitis (Behcet syndrome) and encephalomyelomeningitis. Acta Psychiatr Neurol Scand. 1951;26:443–53.
20. McMenemey WH, Lawrence BJ. Encephalomyelopathy in Behcet disease: report of necropsy findings in two cases. Lancet. 1957;2:353–8.
21. Kawakita H, Nishimura M, Satoh Y, et al. Neurological aspects of Behcet disease: a case report and clinico-pathological review of the literature in Japan. J Neurol Sci. 1967;5:417–39.

22. Arai T, Mizukami K, Sasaki M, et al. Clinicopathological study on a case of neuro-Behcet disease: in special reference to MRI, SPECT and neuropathological findings. Jpn J Psychiatry Neurol. 1994;48:77–84.
23. Hadfield MG, Aydin F, Lippman HR, et al. Neuro-Behcet disease. Clin Neuropathol. 1996;15:249–55.
24. Mirsattari SM, McGinn GJ, Halliday WC, et al. Neuro-Behcet disease with predominant involvement of the brainstem. Neurology. 2004;63:382–4.
25. Yamada H, Saito K, Hokari M, Toru S. Brain biopsy to aid diagnosis of neuro-Behçet's disease: case report and literature review. eNeurologicalSci. 2017;8:2–4.
26. Siva A, Altintas A, Saip S. Behçet's syndrome and the nervous system. Curr Opin Neurol. 2004;17(3):347–57.
27. Kara B, Somer A, Pişkin S, Aydinli N, Salman N, Yalçin I. Neuro-Behçet syndrome presenting as acute meningeal syndrome. J Infect. 2006;52(4):e120–3.
28. Gallucci M, Smith JD, Limbucci N, Rossi A, Demaerel P, Krings T, D'Amico A, Micheli C. Pediatric inflammatory diseases. Part IV: miscellaneous, Reye, PRES, Sarcoidosis. Neuroradiol J. 2012;25(6):725–38.
29. Kari JA, Shah V, Dillon MJ. Behçet's disease in UK children: clinical features and treatment including thalidomide. Rheumatology. 2001;40:933–8.
30. Vesoulis ZA, Galindo R, Ornstein BW, White AJ. Neuro-Behçets in a child. Child Neurol Open. 2014;1(1):2329048X14550505.
31. Uluduz D, Kürtüncü M, Yapıcı Z, Seyahi E, Kasapçopur Ö, Özdoğan H, Saip S, Akman-Demir G, Siva A. Clinical characteristics of pediatric-onset neuro-Behçet disease. Neurology. 2011;77(21):1900–5.
32. Lee SH, Yoon PH, Park SJ, Kim DI. MRI findings in neuro-Behçet's disease. Clin Radiol. 2001;56(6):485–94.
33. Erdem E, Carlier R, Idir AB, Masnou PO, Moulonguet A, Adams D, Doyon D. Gadolinium-enhanced MRI in central nervous system Behçet's disease. Neuroradiology. 1993;35(2):142–4.
34. Morrissey SP, Miller DH, Hermaszewski R, Rudge P, MacManus DG, Kendall B, McDonald WI. Magnetic resonance imaging of the central nervous system in Behçet's disease. Eur Neurol. 1993;33(4):287–93.
35. Akman-Demir G, Bahar S, Coban O, Tasci B, Serdaroglu P. Cranial MRI in Behçet's disease: 134 examinations of 98 patients. Neuroradiology. 2003;45(12):851–9.
36. Nüssel F, Wegmüller H, Laseyras F, Posse S, Herschkowitz N, Huber P. Neuro-Behçet: acute and sequential aspects by MRI and MRS. Eur Neurol. 1991;31(6):399–402.
37. Scully SE, Stebner FC, Yoest SM. Magnetic resonance spectroscopic findings in neuro-Behçet disease. Neurologist. 2004;10(6):323–6.
38. Borhani Haghighi A, Sarhadi S, Farahangiz S. MRI findings of neuro-Behcet's disease. Clin Rheumatol. 2011;30(6):765–70.
39. Alkan A, Goktan A, Karincaoglu Y, Kamisli S, Dogan M, Oztanir N, Turan N, Kocakoc E. Brain perfusion MRI findings in patients with Behcet's disease. Sci World J. 2012;2012:261502.
40. Park JH, Jung MK, Bang CO, Park HK, Sung KB, Ahn MY, et al. Neuro-Behçet's disease mimicking a cerebral tumor: a case report. J Korean Med Sci. 2002;17:718–22.
41. Matsuo K, Yamada K, Nakajima K, Nakagawa M. Neuro-Behçet disease mimicking brain tumor. AJNR Am J Neuroradiol. 2005;26:650–3.
42. Schmolck H. Large thalamic mass due to neuro-Behçet disease. Neurology. 2005;65:436.
43. Heo JH, Lee ST, Chu K, Kim M. Neuro-Behçet's disease mimicking multiple brain tumors: diffusion-weighted MR study and literature review. J Neurol Sci. 2008;264:177–81.
44. Tokgoz OS, Akpinar Z, Guney F, Seyithanoğlu A. A Neuro-Behçet's case operated with the intracranial mass misdiagnosis. J Korean Neurosurg Soc. 2012;52(5):488–90.
45. Bilge NŞ, Şaylısoy S, Kaşifoglu T, Korkmaz C. Mass-like lesions as a rare form of neuro-Behçet's disease: a case report and review of the literature. Eur J Rheumatol. 2014;1(1):34–8.
46. Cohen-Aubart F, Psimaras D, Galanaud D, Bielle F, Touitou V, Haroche J, Amoura Z. Cerebral pseudo-tumoral neuro-Behcet: histological demonstration of an inflammatory and vascular disease. Clin Neurol Neurosurg. 2017;161:48–50.

47. Noel N, Hutié M, Wechsler B, Vignes S, Le Thi Huong-Boutin D, Amoura Z, Dormont D, Delcey V, Polivka M, Cacoub P, Saadoun D. Pseudotumoural presentation of neuro-Behcet's disease: case series and review of literature. Rheumatology (Oxford). 2012;51(7):1216–25.
48. Fonseca Cardoso A, Rocha-Filho PA, Melo Correa-Lima AR. Neuro-Behçet:differential diagnosis of recurrent meningitis. Rev Med Chil. 2013;141(1):114–8.
49. Lamzaf L, Harmouche H, Alaoui-Bennesser H, Mezalek ZT, Adnaoui M, Aouni M. Recurrent meningitis revealing Behçet disease: report of two cases. Pan Afr Med J. 2014;19:14.
50. Devlin T, Gray L, Allen NB, Friedman AH, Tien R, Morgenlander JC. Neuro-Behçet's disease: factors hampering proper diagnosis. Neurology. 1995;45(9):1754–7.
51. Gumà A, Aguilera C, Acebes J, Arruga J, Pons L. Meningeal involvement in Behçet's disease: MRI. Neuroradiology. 1998;40(8):512–5.
52. Borhani Haghighi A, Ashjazadeh N. Acute disseminated encephalomyelitis-like manifestations in a patient with neuro-Behçet disease. Neurologist. 2009;15(5):282–4.
53. Yesilot N, Mutlu M, Gungor O, Baykal B, Serdaroglu P, Akman-Demir G. Clinical characteristics and course of spinal cord involvement in Behçet's disease. Eur J Neurol. 2007;14(7):729–37.
54. Uygunoğlu U, Pasha M, Saip S, Siva A. Recurrent longitudinal extensive transverse myelitis in a neuro-Behçet syndrome treated with infliximab. J Spinal Cord Med. 2015;38(1):111–4.
55. Coban O, Bahar S, Akman-Demir G, Taşci B, Yurdakul S, Yazici H, Serdaroğlu P. Masked assessment of MRI findings: is it possible to differentiate neuro-Behçet's disease from other central nervous system diseases? [corrected]. Neuroradiology. 1999;41(4):255–60.

Neuropsychiatric Symptoms in Neuro-Behçet's Disease

Tuncay Gündüz and Erhan Ertekin

Introduction

The central nervous system (CNS) involvement in Behçet's disease (BD) may cause significant disability in both acute and chronic phases. Involvement of predominantly brainstem and diencephalic structures and less commonly other regions in neuro-Behçet's disease (NBD) patients may explain the substantially debilitating nature of the disease. Among these findings, cognitive involvement and neuropsychiatric involvement hold a special place which can cause radical changes in the patient's private, work, and social life in both the acute and chronic phases. Perhaps these deficits would be the most significant obstacles in the patients' lives.

In general, dysfunction of specific cognitive domains, apparent personality changes, and psychiatric disorders may occur with different severity and combinations in NBD patients. In this chapter, cognitive dysfunction and personality changes are presented first, followed by psychiatric symptoms and findings.

BD is associated with a high prevalence of psychiatric symptoms, and the chronic nature of the disease and its significant morbidity can cause problems in patients' lives as shown by poor quality of life and diminished life satisfaction in patients with BD [8]. The relationship between stress and the immune system also should be taken into account when evaluating the psychological aspects of the disease, since BD is an immuno-inflammatory disease. Approximately 70% of patients with BD reported a stress factor before the occurrence of the disease, while 80% of patients also declared stress in the period of relapse [30].

T. Gündüz (✉)
Department of Neurology, Istanbul Faculty of Medicine, Istanbul University, Istanbul, Turkey

E. Ertekin
Department of Psychiatry, Istanbul Faculty of Medicine, Istanbul University, Istanbul, Turkey

E. Tüzün, M. Kürtüncü (eds.), *Neuro-Behçet's Disease*,
https://doi.org/10.1007/978-3-030-55273-2_11

Therefore, it is important to assess patients with BD in terms of psychopathology not only for a better understanding of the psychological aspects of the disease but also to develop prevention and/or treatment strategies for this less researched side of BD.

Pathophysiology

There is not much study and information about the pathophysiology of cognitive dysfunction in neuro-Behçet's disease patients [52]. However, in the last two decades, magnetic resonance imaging (MRI) and single-photon emission computerized tomography (SPECT) studies and case reports regarding NBD patients have been published, and some neuroanatomical data related to the affected regions have accumulated.

In the acute phase, mainly ponto-mesencephalic region and diencephalic structures are involved as indicated by clinical and MRI findings in neuro-Behçet's patients. During this phase, T2 hyperintense lesions showing partial contrast enhancement are encountered which are, in most cases, distinct in morphology and distribution. Hyperintense lesions generally disappear entirely within several months in most cases; however, unilateral or bilateral atrophy may occur in these regions later on (Fig. 1).

Usually, cognitive dysfunction has a chronic course in NBD patients. However, MRI may be normal or nearly normal in stable and relapse-free periods. It is evident that there is a permanent dysfunction in some relevant brain areas. Therefore, one of the best methods for demonstrating cerebral dysfunction, SPECT, was performed in patients with NBD.

Mizukami et al. found frontal and temporal hypometabolism in two progressive NBD patients with dementia [37]. Arai et al. performed a SPECT, MR, and autopsy study in a 50-year-old NBD patient with a frontal syndrome demonstrating disinhibition, the absence of spontaneity, and memory impairment. They found severe brainstem involvement in MRI and severe frontal hypometabolism in SPECT. However, an autopsy showed gliosis and perivascular infiltration of the brainstem, internal capsule, and thalamus without involvement of the frontal cortex [4].

A SPECT study of Garcia-Burillo and colleagues found perfusion defects mainly in the frontal, temporal, and basal ganglia regions and less commonly in occipital regions. Furthermore, of the 21 patients with no neurological findings, nine had abnormal SPECT findings [23].

Vignola and colleagues reported that, despite normal brain MRI, SPECT study revealed hypoperfusion regions mainly in the basal ganglia and temporal and frontal lobes in five boys with NBD demonstrating various neurological signs and symptoms. They linked these findings to disseminated cerebral inflammation [55].

Huang and colleagues observed mainly parietal, temporal, and frontal hypometabolism in ten NBD patients with predominantly cognitive complaints without brain MR lesions [29].

Mimura detected memory impairment, frontal/executive dysfunction in a patient with amnesia, hypoperfusion in bilateral basal ganglia and basal forebrain, and

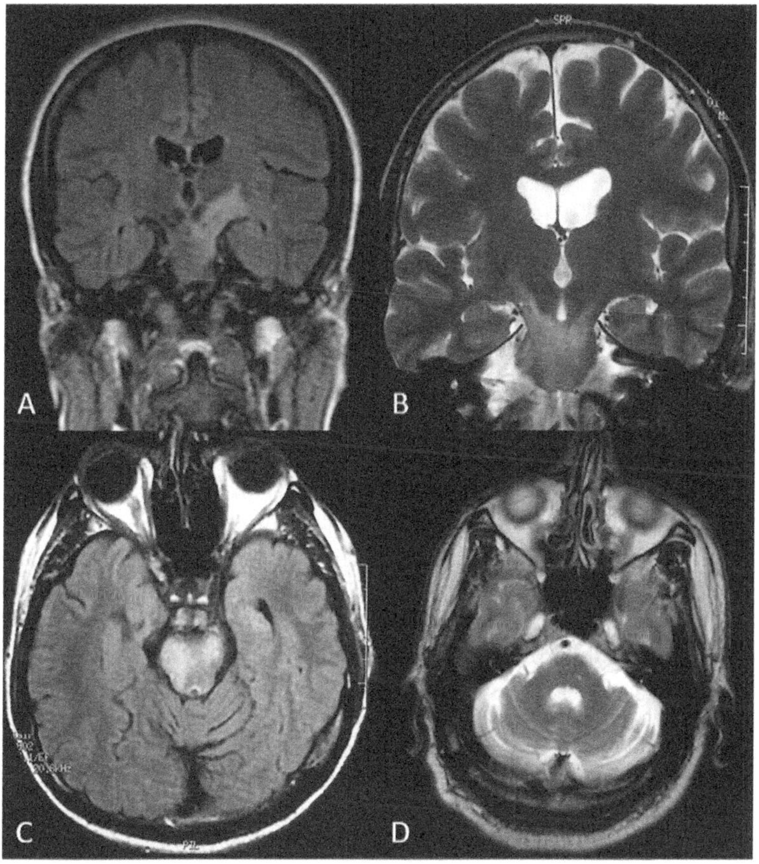

Fig. 1 MRI samples of NBD patients. (**a, b, c**) Acute brainstem lesions with a characteristic appearance. (**d**) In chronic term, prominent atrophy in the brainstem and cerebellum is evident

hyperperfusion in posterior cortical regions. Brain MRI showed a small high intensity region in the left thalamus together with mild dilatation of the third ventricle. The authors linked this deranged neuropsychological profile to dysfunctioning thalamic, subcortical-frontal circuits [35].

Chuang et al. reported that all patients with NBD in their cohort (n:24) showed abnormal SPECT findings in which temporal lobe and basal ganglia were predominantly affected despite normal brain MRI. However, authors did not report neuropsychological profiles of the patients, precluding the opportunity to build a neuroanatomical–cognitive dysfunction correlation [12].

In summary, SPECT studies in NBD patients show that predominantly affected areas are the frontal lobe, temporal lobe, and basal ganglia. In accordance with this, cognitive dysfunction profiles of these patients are more or less consistent with the involvement of these same regions. Interestingly, the brain MRI often can not detect an abnormality in these regions.

Normal MRI or sole brainstem atrophy despite the apparent cognitive dysfunction detected in these patients may suggest that MRI may not detect responsible micro-lesions in these patients [25, 41].

Another question that may come to mind is: why do we detect perfusion abnormalities in normal MRI regions and despite normal histopathological examination?

Frontal and executive functions are very complicated, and their physiology is still waiting to be discovered. On the other hand, some operating mechanisms have been introduced so far. As known, basal ganglia, besides the coordination of the motion, also take a significant role in the regulation of the movement inducer factors (e.g., emotion, motivation, and cognition) [26]. This process is established by complex and interrelated circuits between the basal ganglia neurons and various frontal cortical regions (cortico-striatal circuits). Through these circuits, while particular behaviors are performed outputs are modified appropriately with internal and external stimuli. Also, individual striatal structures exchange extensive information with the brainstem dopaminergic neurons and tegmental nuclei. Naturally, one can expect that damage to the basal ganglia and brainstem may impair frontal functions. Indeed, dementia in parkinsonism, and depression and emotional instability encountered in brainstem involvement may be taken as suitable examples of this proposition [26].

Another modulator of frontal executive functions is considered to be cerebellum. The influence of cerebellum on cognitive functions, which is known more as the movement regulator, has recently been suggested [14, 33]. Cortical-cerebellar interaction depends mainly on two pathways: the afferent cortico-ponto-cerebellar pathway and the efferent cerebello-thalamo-cortical pathway. The afferent fibers reach to cerebellar nuclei through middle cerebellar peduncle of the pons and leave through the super cerebellar peduncle [43]. In their diffusion tensor imaging (DTI) MRI study, Palesi et al. showed that one of the main relay stations of these two pathways was the prefrontal cortex which may reveal the importance of cerebellum in cognition [43] (Figs. 2 and 3).

The cognitive impairment caused by cerebellar diseases is notable. Cerebellar cognitive affective syndrome (CCAS) caused by cerebellar damage was first described in 1990 by Schmahmann and colleagues and the concept evolved ever since. It is characterized by disturbances in executive functions, linguistic processing, spatial cognition, and affect regulation [28, 46]. Likewise, as the heterogeneous literature indicates, common picture resembles frontal lobe dysfunction, with altered working memory, verbal fluency, and cognitive control [48]. However, the cognitive effects of cerebellar damage may not be as evident as cerebral damage because there are a number of different cerebello-cortical interactions complicating the clinical picture. Additionally, the damage will often extend to other brainstem regions which may also complicate the neuropsychological profile [48]. D'Aes and Marien reviewed patients with isolated brainstem infarctions and identified 30 patients with sufficient anatomoclinical information. The main cognitive profile was frontal dysfunction (attention, executive functions, and memory), and SPECT study revealed hypoperfusion mainly in the frontal region, followed by parietal and temporal regions. The authors suggested that the primary cause of cognitive impairment in

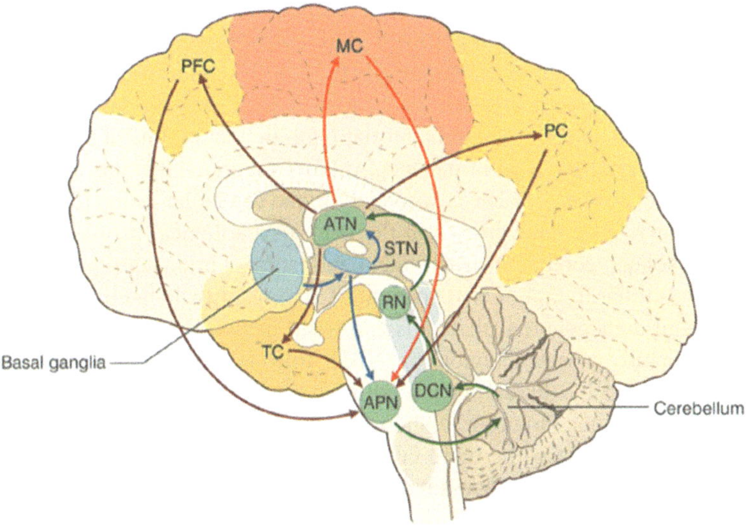

Fig. 2 The cerebello-thalamo-cerebro-cortical circuits (CTCCs). The figure represents schematically the bidirectional connectivity between the cerebellum and the telencephalon, in particular with the cereberal cortex. Telencephalic projections from the cortex and basal ganglia (through the subthalamic nucleus, STN) and limbic areas are relayed to the cerebellum through the anterior pontine nuclei (APN). The cerebellum in turn sends its output through the deep cerebellar nuclei (DCN), red nucleus (RN), and anterior thalamic nucleus (ATN) to various telencephalic areas including the motor cortex (MC), the prefrontal cortex (PFC), the parietal cortex (PC), and the temporal cortex (TC). These connections, which are supported by anatomical and functional data, forming several bidirectional cerebello-thalamo-cerebro-cortical circuits (CTCCs). (Figure is reproduced from D'Angelo et al. Front Neural Circuits. 2013;6:116. https://www.frontiersin.org/articles/10.3389/fncir.2012.00116/full) [14]

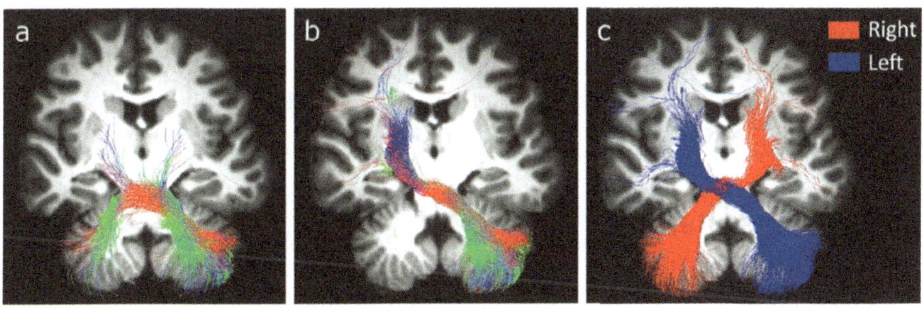

Fig. 3 2D rendering of combined CSD technique and probabilistic streamlines tractography in a representative subject. Streamlines in (**a**, **b**) are color coded according to the diffusion-derived fiber direction, while in (**c**), two solid colors differentiate the left and the right seeded pathways. (**a**) Pathway reconstructed with a seed ROI placed in the left middle cerebellar peduncle. No target ROI was used. (**b**) Pathway reconstructed with a seed ROI in the left middle cerebellar peduncle and a target ROI in the contralateral cerebral peduncle. (**c**) Pathways reconstructed as in (**b**) using both left (blue) and right (red) seed ROIs (Figure is reproduced from Palesi et al. Sci Rep. 2017;7(1):12841. https://www.nature.com/articles/s41598-017-13079-8) [43]

these patients may be the involvement of the cortical-cerebellar pathways which pass through the brainstem and the impairment of the modulatory effect of the cerebellum on higher cortical and limbic functions [13]. Therefore, nonstructural biochemical deterioration may also be involved in the pathological process. Hoffman and colleagues studied 16 patients with subtentorial infarction and found at least one of the frontal network syndromes in 11 patients. Authors suggested that this may be due to the contralateral cortical diaschisis secondary to the cerebellar lesion [56].

In summary, it is noteworthy that the neuropsychological profile and SPECT findings in isolated cerebellar involvement and isolated brainstem involvement are more or less similar to that of NBD patients. Consequently, it can be suggested that brainstem and diencephalon (thalamus and basal ganglia) involvement, which are the predominantly affected regions in NBD, may lead to this distinct cognitive impairment profile by disrupting the cortico-striatal and cortico-cerebellar pathways in these patients. However, it should be kept in mind that cognitive dysfunction in these patients may be more complicated and mixed clinical pictures may emerge as the pathological process may involve more extensive structures.

Prevalence

Cognitive dysfunction has been reported in NBD patients with a variable prevalence from 30% to as high as 100%. Characteristics of clinical presentations will be discussed below. As mentioned above, prevalence of psychiatric disorders in patients with NBD has not been thoroughly investigated by using structured interviews. In the only study that assessed patients with BD to formally diagnose psychiatric disorders, Dursun et al. recruited 73 patients and assessed them with Structured Clinical Interview for DSM-IV/Clinical Version (SCID-I/CV). Thirty patients (41.1%) reported at least one current psychiatric disorder, and anxiety disorders were more common than mood disorders (35.6% and 21.9%, respectively). Major depression (17.8%) was the most frequent psychiatric disorder, and it was followed by specific phobia (16.4%), generalized anxiety disorder (15.1%), and social phobia (9.6%) [17].

Clinical Presentations

There are not many studies concerning cognitive impairment in neuro-Behçet's disease. However, our experience and few published studies and case reports indicate that primarily memory, attention, and executive dysfunction are observed in these patients [2, 11, 24, 25, 41]. On the other hand, other cognitive domains may be affected to some extent in individual patients. Personality changes may be observed with or without dysfunction in these cognitive domains.

Neuropsychological evaluations (NPEs) of these patients are usually performed after an acute exacerbation subsided and the patient has been stabilized. Some acute exacerbations may require hospitalization and intense immunosuppressive treatment.

Symptoms	Prevalence
Secondary memory disorder (especially delayed recall)	30–100%
Frontal executive dysfunction Impaired learning, impaired self-care, disinhibition, perseverations, apathy Failure in persistence, category shifting, response inhibition, and mental control Pseudobulbar affect	60%
Depression and anxiety	N/A
Hypomania/mania and psychosis	N/A
Kleptomania	N/A

N/A: Data not available in neuro-Behçet's disease patients

Memory Disorders

Temporolimbic recording process is the mainstay for the acquisition of new information and memory. Deterioration of this process, named primary memory disorder, is a common finding in some disorders such as Alzheimer's disease which predominantly affect temporolimbic structures in the early phases. On the other hand, stored information should be retrieved into consciousness either spontaneously/actively (retrieval) or passively by clue (recognition). These processes require intact attention and other complex frontal functions to be operational. Therefore, patients with these frontal dysfunctions cannot remember spontaneously during testing but remember when given a clue (recognition), hence named as secondary memory disorders.

Neuro-Behçet's disease patients frequently complain of forgetfulness. Memory profile of affected patients indicates almost always a secondary memory disorder, i.e., secondary to frontal dysfunction. These patients demonstrate intact registration but impaired retrieval. However, in a small group of patients in which complex attention functions deteriorated severely, recording defect, i.e., primary memory disorder, is also seen.

The literature review also reveals more or less the same pattern of cognitive involvement in these patients. In our center's cohort, all evaluated NBD patients (74 patients) showed a memory disorder of which 70% was medium-severe [2]. On the other hand, memory scores showed substantial improvement with recognition. Also, 12 patients' consecutive NPE revealed preserved registration and storage but significant impairment of delayed recall, especially in visual and verbal modalities [41]. Memory scores of these patients also improved with recognition trials dramatically. Cavaco et al. found short-term memory and immediate recall disturbance with attention deficit in 15 patients [11], while Gökçay et al. reported immediate recall dysfunction in 30% of NBD patients [24].

We previously reported that memory and attention deficits were identified in both NBD and multiple sclerosis patients, although the findings related to prefrontal involvement were more prevalent in NBD patients. Learning and early and delayed recall scores were worse in NBD patients, but recognition was preserved [25].

Attention and Frontal Executive Dysfunctions

Neuro-Behçet's disease patients may give a clinical impression that they are easily distracted and lacking self-care. Indeed, in the vast majority of patients, it appears that there is deterioration at various levels in attention processes. Likewise, impaired frontal executive functions such as deterioration in response inhibition, and perseverations with variable severity may be observed. Patients may have long-lasting apathy, depression, pseudobulbar affect, and disinhibited behaviors influencing every aspect of the social and personal life. Some rare patients may exhibit kleptomania behavior. Relatives and household of patients may report a personality change, imperviousness, indifference, inadequate self-care, and disruption in social life and relationships.

Studies indicate that attention deficit and frontal executive dysfunction may be evident in as much as 60% of patients [2]. Oktem-Tanör's study revealed a 60% clinical impression of personality change, 50% displayed disinhibition, and 15% displayed apathy. Five patients showed failure in persistence, category shifting, response inhibition, and mental control [41]. Comparison of NBD and multiple sclerosis (MS) patients revealed worse learning strategies (less semantic clustering) and worse performance in Stroop test and frontal behavioral inventory in NBD patients. Overall, these findings could be attributed to the more severe involvement of the fronto-striatal circuits in NBD as compared to MS [25].

Cognitive Involvement in Other Domains

Visuospatial functions, which are regulated mainly in the posterior cerebral regions, appear to be preserved in general. However, patients may not perform well during testing, often due to severe attention deficit. Patients' language functions are generally preserved, and they generally do not have a problem other than dysarthria.

Cognitive Dysfunction in Patients Without Apparent Neurological Involvement

Another proposed pattern is insidious cognitive impairment in patients without any neurological symptoms. Conflicting results have been reported in several studies. One study reported decreased cognitive event-related potential (p300) in Behçet's patients without neurological involvement [31]. In another study, 26 BD patients were compared to healthy subjects. They found deficits in memory, executive functions, and attention in 46% of BD patients. However, the lack of MR imaging data makes the findings doubtful [38]. On the contrary, there are studies that found no difference in NPE and p300 latency values between Behçet's disease patients and healthy subjects [24]. In conclusion, current evidence does not prove the presence of subclinical neurological/cognitive involvement in Behçet's disease patients without neurological involvement.

Depression and Anxiety Symptoms

Most studies on psychiatric aspects of BD have been conducted by using rating scales that assess psychiatric symptoms, instead of using structured clinical interviews to formally diagnose psychiatric disorders. In one of the largest of such studies, Bagheri and colleagues evaluated 101 randomly selected patients with BD in Iran with the Symptom Checklist-90-Revised (SCL-90-R), which is a widely used instrument to assess psychiatric symptoms. They found that somatization (91.7%), anxiety (78%), and depression (77.8%) were the most prevalent symptoms [7]. High incidences of these psychiatric symptoms were comparable to the ones reported in much earlier reports on psychiatric profile of patients with BD [20].

Taner and colleagues compared 120 outpatients diagnosed with BD and 95 outpatients with chronic plaque-type psoriasis in terms of depression, anxiety, hopelessness, and automatic thoughts [49]. The authors found that all four scores were significantly higher in the BD group than in the psoriasis group. When they define depression with a Beck Depression Inventory score above 12, they found that 45.5% of patients with BD were depressed and this rate was significantly higher than that of patients with psoriasis as well. They also found a higher rate of depression in BD patients with a disease duration of more than 3 years compared to BD patients with a shorter duration of illness (75% vs. 20%). In the younger BD group (aged 18–25 years), the only score which was higher than that in the psoriasis group was anxiety. In the light of these findings, the authors suggested that anxiety might be the presenting psychiatric symptom in patients with BD, unlike depression which might develop later in the course. In patients with BD, men had significantly higher rates of depression (57%) than women (36.1%). This finding, which is in contrast to the general population where females are more likely to develop depression than males, might be a reflection of the worse prognosis of BD in men [50].

In general, Taner et al.'s findings were comparable to that of an earlier study with a similar design that used a smaller sample in which patients with BD displayed more symptoms of depression and anxiety than both healthy controls and patients with psoriasis [10]. Another study that included 25 patients with BD also reported higher depression and anxiety symptoms in the BD group than in the healthy control group [30]. Other studies confirmed higher depression and anxiety symptoms in patients with BD compared to healthy controls [51] or patients with rheumatoid arthritis, which is also an autoimmune disease [34].

Hypomanic/Manic and Psychotic Symptoms

Compared to the depressive and anxious symptoms, less is known about psychotic and/or manic symptoms in patients with BD. The limited literature on psychosis or bipolar disorder in BD is mostly based on case reports.

Van Ham and colleagues described the case of a patient diagnosed with Behçet's disease after being admitted to a psychiatric hospital with mania. Despite an atypical presentation with no mood elevation or grandiosity, the patient met DSM criteria for a manic episode. The patient had neuro-Behçet's disease with signs of parenchymal involvement. The authors presumed that the patient's psychiatric condition might have been caused by parenchymal brain damage [54]. As reports on the relationship between neuroinflammatory processes and psychiatric disorders are growing in numbers, anti-inflammatory treatment leading to normalization of mood and behavior in 3 months' time is a remarkable data from this case report. Aydin and colleagues reported a case of neuro-Behçet's syndrome with the initial onset of affective symptoms. The patient had a large lesion in the pons and was successfully treated with corticosteroids [6]. Nakano et al. reported a case whose neuro-Behçet's disease emerged with the onset of bipolar disorder [39]. A case report described a patient who developed treatment-resistant bipolar disorder precipitated by BD [3]. A more recent case report described a patient who developed bipolar disorder as the first presentation of the subsequent NBD [27].

Psychotic symptoms have also been reported in patients with BD. Nkam and Cotterau reported an acute psychotic episode in a patient with neuro-Behçet's disease without parenchymal cerebral involvement [40]. Deniz et al. described a patient with neuro-Behçet's disease who was diagnosed as psychotic disorder due to a medical condition and suggested that neuro-Behçet's disease should be considered in psychotic episodes when an organic etiology is suspected [15].

Recently, Otsuka and colleagues reported a case who presented with acute agitation as the initial manifestation of NBD [42].

Clinical Course of Cognitive Dysfunction in Neuro-Behçet's Disease

Usually, the first presentations of neuro-Behçet's disease are brainstem, cerebellar, sensory, motor, and other rare findings depending on the involved brain region. It is known that cognitive impairment is often present from the beginning but may improve, remain same, or worsen in time. On the other hand, in a small group of patients, cognitive symptoms and signs may be the presenting clinical findings [6, 15, 22, 35, 47].

Although there is not much information about the time course of cognitive impairment, most of the patients' conditions will worsen over time. Recurrent neuropsychological evaluation of 12 patients by Oktem-Tanör et al. showed that the memory performance of seven patients (three with relapse and four without relapse) had worsened over time. Other four patients remained unchanged and one patient improved. Attention, frontal functions, and other modalities were not changed in time [41].

The Relationship Between Cognitive Impairment and Psychiatric Symptoms

Cognitive impairment was investigated in a study that enrolled 26 patients with BD and 26 healthy controls. The groups were evaluated with a comprehensive neuropsychological battery as well as the Hamilton scales for depression and anxiety. In line with the literature reviewed above, patients with BD showed significantly higher scores than controls on anxiety and depression scores. Cognitively impaired patients with BD were similar to cognitively unimpaired BD subjects in relation to anxiety and depression levels [38]. However, anxiety might be a risk factor for cognitive impairment in BD. Dutra and colleagues have shown that cognitive impairment occurs frequently in patients with BD independently of neurological manifestation. In that study, low educational level and anxiety emerged as risk factors for cognitive impairment in BD [18].

The Activity of Behçet's Disease and Psychiatric Symptoms

In a cross-sectional study, Abdelraheem et al. compared BD with systemic lupus erythematosus (SLE) in terms of frequency of psychiatric disorders and brain MRI findings. They included 50 patients with SLE, 34 patients with BD, and 44 control subjects. As expected, psychiatric disorders were more prevalent in both rheumatological disorders than healthy controls. However, the prevalence of psychiatric disorders was higher in SLE patients than in BD patients as well (56% and 26.5%, respectively). Although depression was more common in SLE patients than in BD patients, no significant difference in the prevalence of anxiety was found between the two patient groups. Among patients with psychiatric manifestations, significantly more MRI abnormalities were found in BD patients than in SLE patients. This might indicate a close association between neuro-Behçet's disease and psychiatric manifestations.

The same study also assessed disease activity for both disorders and reported that major psychiatric conditions like psychosis and mania might be associated with higher disease activity, whereas the groups with less disease activity showed significant accumulation of depression and anxiety. The authors suggested that this finding might be linked to the more chance of cerebral vasculitis in higher disease activity rendering patients with higher propensity to have more aggressive psychiatric disorders [1].

Neuropsychiatric Symptoms Secondary to Medications

For azathioprine, a case with confusion and another with obsessive-compulsive disorder and panic disorder were reported as possible side effects [53, 54]. Likewise, transient mood disorders secondary to mycophenolate mofetil [5, 16] and psychotic symptoms after tacrolimus use [9] have been reported. Kenna et al. reported that complications of corticosteroid treatment may cause neuropsychiatric symptoms ranging from significant anxiety and insomnia to severe mood and psychotic disorders, delirium, and dementia [32]. Infliximab use has been reported to cause mood disorders as manic switch and severe depression leading to suicide attempt [19, 21].

Management

No study was conducted to investigate the treatment of cognitive symptoms in neuro-Behçet's patients. However, psychiatric symptoms and signs of the patients can be controlled by antidepressants, anxiolytics, and antipsychotics. Cognitive rehabilitation is likely to work, but no studies have been done on it. Various rehabilitation studies have been carried out for cognitive impairment in multiple sclerosis and other chronic CNS disorders. Some of these revealed a lack of efficacy [36], while some provided low-level evidence of a positive effect [45]. However, it has also been suggested that cognitive neurorehabilitation may have a positive effect on isolated brainstem infarcts [13].

Evidence is also very limited regarding the treatment of psychiatric symptoms in patients with BD. The psychotic symptoms of the patient reported by Deniz et al. did not improve after steroid treatment, and later, they were partially reduced after the addition of risperidone [15]. Patel et al. reported an adolescent patient with BD presented with psychosis who showed improvement with a combination of corticosteroids, anticoagulants, and immunosuppressants [44]. Due to limited evidence, it is not possible to determine the best treatment for psychiatric symptoms in patients with BD. The optimal treatment might be corticosteroids and other immunosuppressant drugs for some patients, whereas other patients might require psychotropic medications such as antidepressants, antipsychotics, and mood stabilizers.

References

1. Abdelraheem T, Habib HM, Eissa AA, Radwan NM. Psychiatric disorders and MRI brain findings in patients with systemic lupus erythematosus and Behcet's disease: a cross sectional study. Acta Reumatol Port. 2013;38(4):252–60.
2. Akman-Demir G l, Serdaroglu P, Taççı B. Clinical patterns of neurological involvement in Behcet's disease: evaluation of 200 patients. Brain. 1999;122(11):2171–81.

3. Alevizos B, Anagnostara C, Christodoulou GN. Resistant bipolar disorder precipitated by Behcet's syndrome. Bipolar Disord. 2004;6(3):260–3.
4. Arai T, Mizukami K, Sasaki M, Tanaka Y, Shiraishi H, Horiguchi H, Ogata T. Clinicopathological study on a case of neuro-Behcet's disease: in special reference to MRI, SPECT and neuropathological findings. Jpn J Psychiatry Neurol. 1994;48(1):77–84.
5. Arkin L, Talasila S, Paller AS. Mycophenolate Mofetil and mood changes in children with skin disorders. Pediatr Dermatol. 2016;33(3):e216–7.
6. Aydin N, Aydin MD, Deniz O, Kirpinar I. Neuro-Behcet's disease involving the pons with initial onset of affective symptoms. Eur Arch Psychiatry Clin Neurosci. 2002;252(1):44–6.
7. Bagheri F, Mani A, Tadayyoni A, Firozi F, Nazarinia MA. The prevalence of psychiatric symptoms in the patients with Behçet's disease in shiraz, southwest of Iran. J Mood Disord. 2013;3(1):28–32.
8. Bodur H, Borman P, Ozdemir Y, Atan C, Kural G. Quality of life and life satisfaction in patients with Behcet's disease: relationship with disease activity. Clin Rheumatol. 2006;25(3):329–33.
9. Bourgeois JA, Hategan A. Immunosuppressant-associated neurotoxicity responding to olanzapine. Case Rep Psychiatry. 2014;2014:250472.
10. Calikoglu E, Onder M, Cosar B, Candansayar S. Depression, anxiety levels and general psychological profile in Behcet's disease. Dermatology. 2001;203(3):238–40.
11. Cavaco S, Da Silva AM, Pinto P, Coutinho E, Santos E, Bettencourt A, Pinto C, Gonçalves A, Silva S, Gomes F, Carvalho L, Pereira C, Martins B, Correia J, Vasconcelos C. Cognitive functioning in Behçet's disease. Ann N Y Acad Sci. 2009;1173(1):217–26.
12. Chuang YW, Hsu CC, Huang YF, Lin HF, Chang CC, Lin CY, Tyan YC. Brain perfusion SPECT in patients with Behcet's disease. J Neuroradiol. 2013;40(4):288–93.
13. D'Aes T, Marien P. Cognitive and affective disturbances following focal brainstem lesions: a review and report of three cases. Cerebellum. 2015;14(3):317–40.
14. D'Angelo E, Casali S. Seeking a unified framework for cerebellar function and dysfunction: from circuit operations to cognition. Front Neural Circuits. 2012;6:116.
15. Deniz O, Caykoylu A, Vural G, Albayrak Y, Temel S, Aydin I, Kuloglu M. A case study of neuro-psycho-Behcet's syndrome presenting with psychotic attack. Clin Neurol Neurosurg. 2009;111(10):877–9.
16. Draper HM. Depressive disorder associated with mycophenolate mofetil. Pharmacotherapy. 2008;28(1):136–9.
17. Dursun R, Uguz F, Kaya N, Savas Cilli A, Endogru H. Psychiatric disorders in patients with Behcet's disease. Int J Psychiatry Clin Pract. 2007;11(1):16–20.
18. Dutra LA, de Souza AW, Alessi H, Guedes Bde V, Braga-Neto P, Pedroso JL, Goncalves CR, da Rocha AJ, Bertolucci PH, Barsottini OG. Cognitive impairment in Brazilian patients with Behcet's disease occurs independently of neurologic manifestation. J Neurol Sci. 2013;327(1–2):1–5.
19. Elisa B, Beny L. Induction of manic switch by the tumour necrosis factor-alpha antagonist infliximab. Psychiatry Clin Neurosci. 2010;64(4):442–3.
20. Epstein RS, Cummings NA, Sherwood EB, Bergsma DR. Psychiatric aspects of Behcet's syndrome. J Psychosom Res. 1970;14(2):161–72.
21. Eshuis EJ, Magnin KM, Stokkers PC, Bemelman WA, Bartelsman J. Suicide attempt in ulcerative colitis patient after 4 months of infliximab therapy--a case report. J Crohns Colitis. 2010;4(5):591–3.
22. Fisher CA, Sewell K, Baker A. Chronic behavior disturbance and neurocognitive deficits in neuro-Behcet's disease: a case study. Neurocase. 2016;22(3):332–8.
23. Garcia-Burillo A, Castell J, Fraile M, Jacas C, Vilardell M, Ortega D, Bosch JA. Technetium-99m-HMPAO brain SPECT in Behcet's disease. J Nucl Med. 1998;39(6):950–4.
24. Gökçay F, Celebisoy N, Kisabay A, Kumral E, Akyürekli O. P300 and neuropsychological evaluation in Behçet's disease with and without neurological manifestations. J Neurol. 2004;251(6):676–9.
25. Gunduz T, Emir O, Kurtuncu M, Mutlu M, Tumac A, Akca S, Coban O, Bahar S, Oktem-Tanor O, Tuzun E, Eraksoy M, Gurvit H, Akman-Demir G. Cognitive impairment in neuro-Behcet's disease and multiple sclerosis: a comparative study. Int J Neurosci. 2012;122(11):650–6.

26. Haber SN. Corticostriatal circuitry. Dialogues Clin Neurosci. 2016;18(1):7–21.
27. Hibberd O, Bhomra P, Ahmad A. Radiological neuro-Behçet's associated with bipolar disorder: first presentation of a multisystem disease. J R Coll Physicians Edinb. 2018;48(4):323–7.
28. Hoche F, Guell X, Vangel MG, Sherman JC, Schmahmann JD. The cerebellar cognitive affective/Schmahmann syndrome scale. Brain. 2018;141(1):248–70.
29. Huang WS, Chiu PY, Kao A, Tsai CH, Lee CC. Decreased cerebral blood flow in neuro-Behcet's syndrome patients with neuropsychiatric manifestations and normal magnetic resonance imaging--a preliminary report. J Neuroimaging. 2002;12(4):355–9.
30. Karlidag R, Unal S, Evereklioglu C, Sipahi B, Er H, Yologlu S. Stressful life events, anxiety, depression and coping mechanisms in patients with Behcet's disease. J Eur Acad Dermatol Venereol. 2003;17(6):670–5.
31. Kececi H, Akyol M. P300 in Behcet's patients without neurological manifestations. J Can Sci Neurol. 2001;28(1):66–9.
32. Kenna HA, Poon AW, de los Angeles CP, Koran LM. Psychiatric complications of treatment with corticosteroids: review with case report. Psychiatry Clin Neurosci. 2011;65(6):549–60.
33. Koziol LF, Budding D, Andreasen N, D'Arrigo S, Bulgheroni S, Imamizu H, Ito M, Manto M, Marvel C, Parker K, Pezzulo G, Ramnani N, Riva D, Schmahmann J, Vandervert L, Yamazaki T. Consensus paper: the cerebellum's role in movement and cognition. Cerebellum. 2014;13(1):151–77.
34. Melikoglu MA, Melikoglu M. The relationship between disease activity and depression in patients with Behcet disease and rheumatoid arthritis. Rheumatol Int. 2010;30(7):941–6.
35. Mimura M, Kato M, Kashima H. Neuro-Behcet's disease presenting with amnesia and frontal dysfunction. Clin Neurol Neurosurg. 2009;111(10):889–92.
36. Mitolo M, Venneri A, Wilkinson ID, Sharrack B. Cognitive rehabilitation in multiple sclerosis: a systematic review. J Neurol Sci. 2015;354(1–2):1–9.
37. Mizukami K, Shiraishi H, Tanaka Y, Terashima Y, Kawai N, Baba A, Arai T, Koizumi J. CNS changes in neuro-Behcet's disease: CT, MR, and SPECT findings. Comput Med Imaging Graph. 1992;16(6):401–6.
38. Monastero R, Camarda C, Pipia C, Lopez G, Camarda LK, Baiamonte V, Ferrante A, Triolo G, Camarda R. Cognitive impairment in Behcet's disease patients without overt neurological involvement. J Neurol Sci. 2004;220(1–2):99–104.
39. Nakano Y, Hatanaka Y, Ikebuchi E, Shimizu T, Nanko S, Utsumii T. A case of Neuro-Behcet's disease with early onset of bipolar mood disorder. Seishin Shinkeigaku Zasshi. 2004;106(5):564–73.
40. Nkam I, Cottereau MJ. Acute psychosis and Behcet's disease: a case report. Encéphale. 2006;32(3 Pt 1):385–8.
41. Oktem-Tanör O, Baykan-Kurt B, Gürvit IH, Akman-Demir G, Serdaroğlu P, Öktem-Tanör Ö, Baykan-Kurt B, Gürvit IH, Akman-Demir G, Serdaroğlu P. Neuropsychological follow-up of 12 patients with neuro-Behçet disease. J Neurol. 1999;246(2):113–9.
42. Otsuka Y, Yumoto T, Ihoriya H, Matsumoto N, Sato K, Abe K, Naito H, Nakao A. Acute agitation as an initial manifestation of neuro-Behçet's disease. Case Rep Emerg Med. 2018;2018:5437027.
43. Palesi F, De Rinaldis A, Castellazzi G, Calamante F, Muhlert N, Chard D, Tournier JD, Magenes G, D'Angelo E, Gandini Wheeler-Kingshott CAM. Contralateral cortico-ponto-cerebellar pathways reconstruction in humans in vivo: implications for reciprocal cerebro-cerebellar structural connectivity in motor and non-motor areas. Sci Rep. 2017;7(1):12841.
44. Patel P, Steinschneider M, Boneparth A, Lantos G. Neuro-Behcet disease presenting with acute psychosis in an adolescent. J Child Neurol. 2014;29(9):86–91.
45. Rosti-Otajärvi EM, Hämäläinen PI. Neuropsychological rehabilitation for multiple sclerosis. Cochrane Database Syst Rev. 2014;(2):CD009131. https://doi.org/10.1002/14651858.cd009131.pub3.
46. Schmahmann JD. The cerebellum and cognition. Neurosci Lett. 2018;688:62–75.

47. Shrestha R, Erfan A. Behavioral and cognitive impairments as presenting features of Behcet's disease. J Neuropsychiatry Clin Neurosci. 2013;25(4):E05–7.
48. Sokolov AA, Miall RC, Ivry RB. The cerebellum: adaptive prediction for movement and cognition. Trends Cogn Sci. 2017;21(5):313–32.
49. Taner E, Coşar B, Burhanoğlu S, Çalıkoğlu E, Önder M, Arıkan Z, Calikoğlu E, Onder M, Arikan Z. Depression and anxiety in patients with Behçet's disease compared with that in patients with psoriasis. Int J Dermatol. 2007;46(11):1118–24.
50. Tursen U, Gurler A, Boyvat A. Evaluation of clinical findings according to sex in 2313 Turkish patients with Behcet's disease. Int J Dermatol. 2003;42(5):346–51.
51. Uguz F, Dursun R, Kaya N, Cilli AS. Quality of life in patients with Behcet's disease: the impact of major depression. Gen Hosp Psychiatry. 2007;29(1):21–4.
52. Uygunoglu U, Siva A. Behcet's syndrome and nervous system involvement. Curr Neurol Neurosci Rep. 2018;18(7):35.
53. van der Hoeven J, Duyx J, de Langen JJ, van Royen A. Probable psychiatric side effects of azathioprine. Psychosom Med. 2005;67(3):508.
54. van Ham C, Schrijvers D, De Picker L, Vandendriessche F, Sabbe B. Neuropsychiatric features in Behcet's disease: a case report. Clin Neurol Neurosurg. 2014;127:13–4.
55. Vignola S, Nobili F, Picco P, Gattorno M, Buoncompagni A, Vitali P, Mariani G, Rodriguez G. Brain perfusion spect in juvenile neuro-Behcet's disease. J Nucl Med. 2001;42(8):1151–7.
56. Hoffmann M, Cases LB. Etiology of frontal network syndromes in isolated subtentorial stroke. Behav Neurol. 2008;20(3):101–105. https://doi.org/10.3233/BEN-2008-0220

Treatment of Neuro-Behçet's Disease

Murat Kürtüncü, Burcu Altunrende, and Gülşen Akman

Introduction

Neuro-Behçet's disease (NBD) is the most severe complication of Behçet's disease (BD) that requires effective management strategies to prevent long-term disability. However, due to the lack of placebo-controlled treatment trials, NBD is usually treated empirically based on studies on other immune-mediated central nervous system (CNS) diseases, as well as other systemic involvement types of BD. As expected, not all conclusions from other system involvements of BD can be extrapolated to patients with NBD. For example, cyclosporine, which is commonly used for uveitis, may provoke cerebral parenchymal lesions in otherwise neurologically asymptomatic patients and should be contraindicated in patients with NBD [1, 2].

The aim of the treatment must be focused on rapid suppression of acute inflammation and prevention of further relapses by using an aggressive management strategy depending on the clinical severity of the involvement. Acute inflammation should be treated as early as possible with high-dose intravenous (i.v.) corticosteroids followed by immunosuppressive agents with prolonged corticosteroid maintenance, depending on the recovery. Oral immunosuppressant drugs should be swiftly escalated to more effective treatment regimens such as cyclophosphamide (CYC) and anti-TNF-α monoclonal antibodies in relapsing and progressive patients.

M. Kürtüncü (✉)
Department of Neurology, Istanbul Faculty of Medicine, Istanbul University, Istanbul, Turkey

B. Altunrende · G. Akman
Department of Neurology, Demiroğlu Bilim University, Istanbul, Turkey

E. Tüzün, M. Kürtüncü (eds.), *Neuro-Behçet's Disease*,
https://doi.org/10.1007/978-3-030-55273-2_12

Treatment of Acute Relapse

The mainstay of treatment of an acute relapse is high-dose (1000 mg/day) intravenous (i.v.) methylprednisolone (MP), which is also the first treatment option in various immune-mediated CNS diseases [3]. The favorable effect of a 5–10-day MP infusion can usually be seen in 1–4 weeks after treatment. MP treatment should be followed by a slow oral steroid taper over several months, depending on the clinical severity and pattern of involvement.

At the Istanbul University, our conventional treatment protocol consists of oral MP 32 mg/day for at least 4 weeks with intermittent weekly MP 1000 mg i.v. treatments. Oral MP may be gradually decreased to a maintenance dose of 8–16 mg/day after 3 months. The biological effects of disease-modifying treatments might not peak for as long as 3–6 months after initiation. Therefore, we recommend continuing higher doses of oral MP for a total of 3–6 months for cerebral venous sinus thrombosis (CVST) and aseptic meningitis and 6–12 months for parenchymal NBD (p-NBD).

Side effects of prolonged corticosteroid use are gastric ulcers, osteoporosis, hypertension, diabetes mellitus, Cushing's syndrome, psychiatric disorders, and steroid myopathy, which may mimic exacerbation of BD.

The prophylactic effect of MP treatment in acute NBD attacks is still uncertain. However, MP given at regular intervals may have a beneficial effect in the long term, especially according to studies conducted in other CNS autoimmune diseases, such as multiple sclerosis (MS) [4].

Intractable cases may be treated with i.v. cyclophosphamide (CYC) or IVIG. (CYC) is a potent immunosuppressive agent that will be discussed in detail. Due to its anti-inflammatory activity in various autoimmune diseases, i.v. immunoglobulin (IVIG) has been proven to be effective in many forms of central and peripheral neurological disorders. Unfortunately, the effect of IVIG on BD is uncertain. Despite the absence of randomized clinical trials (RCTs) on IVIG, there are case reports that IVIG may be beneficial in the prophylactic treatment of patients with resistant ocular [5], cutaneous, and even neurological involvements [6]. Therefore, it may be a treatment alternative in NBD that can only be used for patients with contraindications to corticosteroids.

Prevention of Relapses

Basic Concepts

The rate of disability accumulation of relapsing or progressive patients with NBD is high, especially in patients with p-NBD. Additionally, relapses in p-NBD occur almost twice as common as CVST [7]. Therefore, all patients should be advised to take disease-modifying treatment after the first relapse. However, since there are only scarcely any guidelines based on evidence-based medicine, clinicians may undergo challenging clinical decisions, principally in refractory patients with NBD.

Attacks and progression in NBD can be prevented by using first-line immunosuppressive agents such as azathioprine, mycophenolate, and methotrexate (Table 1). Azathioprine is usually the primary drug of choice, partly due to the cost and ease of availability. The immunosuppressants should be started as soon as possible after the first clinical attack since their full biological effect may be delayed for up to 3–6 months. During that period, patients should be kept on corticosteroid maintenance treatment. Other than azathioprine, mycophenolate mofetil can also be used as a first-line agent. Another first-line treatment alternative is methotrexate, although there is a scarcity of scientific evidence on its use in NBD except for a few case reports [8, 9]. It is advised to be used only in patients who cannot tolerate azathioprine and mycophenolate.

Second-line treatments should be used in refractory patients. These agents include anti-TNF-α drugs and cyclophosphamide. In the last 5 years, anti-TNF-α agents have become more popular due to their favorable side effect profile in the long term in comparison to cyclophosphamide (Table 1). Despite the absence of high-grade scientific data, interferon-α can also be included in this group.

Patients refractory to second-line treatments may receive anti-IL-1 or anti-IL-6 monoclonal antibodies. However, there is insufficient evidence on these

Table 1 Treatment algorithm of neuro-Behçet's disease

Acute relapse	First-line	Second-line	Third-line
Methylprednisolone (MP): 1 g i.v. 5–10 days ± weekly pulses+ p.o. MP with very slow taper over months up to 1 year	*Azathioprine*: 2.5 mg/kg/day p.o.	*Cyclophosphamide*: 1000 mg/month i.v. 6–12 months	*Anakinra*: 100 mg/day s.c.
Cyclophosphamide: 1000 mg i.v. infusion or 700 mg/m² i.v. infusion	*Mycophenolate mofetil*: 2 gr/day p.o.	*Infliximab*: 5 mg/kg i.v. 0, 2., 6. weeks than every 2 months	*Canakinumab*: 300 mg i.v. loading followed by 150 mg i.v. every month. Followed by 150 mg s.c. every month
	Methotrexate: 12.5–25 mg/week p.o.	*Etanercept*: 25 mg sc. two times per week or 50 mg/week	*Tocilizumab*: 4–8 mg/kg i.v. infusion every month
		Adalimumab: 40 mg s.c. every 2 weeks	*IFN-α*: 6–9 MIU/day s.c. for 1 week and tapered down to 3 MIU s.c. three times a week
			High dose chemotherapy with autologous stem cell transplantation

MP methylprednisolone, *i.v.* intravenous, *s.c.* subcutaneous, *p.o.* oral medications in patients with NBD. Interferon-α treatments may also be tried on patients. Even though there is some evidence that suggests the efficacy of interferon in patients with NBD, the probability of responding to interferon-α in a patient who is refractory to second-line agents is very low. Therefore, even when tried, interferon must be given as an add-on treatment in this clinical situation. Autologous hematopoietic stem cell transplantation (HSCT) with high-dose chemotherapy is a relatively novel treatment option in refractory patients with BD. There are solely case reports of patients who responded at least partially to HSCT (Table 1).

Anticoagulation for Cerebral Venous Sinus Thrombosis

The addition of anticoagulants to steroids is not routinely performed since it is still controversial. There are two reasons for this controversy. First, BD patients with CVST are more likely to have systemic large vessel disease, including pulmonary and peripheral aneurysms that carry a high risk of rupture and bleeding. Second, the suppression of inflammation by immunosuppressants usually recanalizes the thrombotic vessel caused by focal inflammatory cell infiltrate, per se. Therefore, it is advised to reserve anticoagulants only for patients who are refractory to corticosteroids and immunosuppressants as an add-on option [3, 10].

Drug Treatments for NBD

Azathioprine

Azathioprine is a purine analog that is converted to its active metabolite, 6-mercaptopurine (6-MP). It exerts an immunosuppressive effect by inhibition of de novo synthesis of purine nucleotides. Azathioprine has been recommended to use in the treatment of mucocutaneous lesions, eye and joint involvement (level of evidence: IB, strength of recommendation: A), deep vein thrombosis (level of evidence: III, strength of recommendation: C), of BD, and p-NBD (level of evidence: III, strength of recommendation: C) [11]. There are no RCTs evaluating its efficacy in p-NBD or CVST. There is only a single randomized, double-blind, placebo-controlled trial on its efficacy in patients with BD [12]. Another study also showed a beneficial effect in the long term [13]. Although these clinical trials cannot accomplish a decisive conclusion on the effect of NBD, observational analysis of a large number of cases with NBD suggests that azathioprine can control disease activity regardless of disease duration [14]. Although there are no RCTs, according to observational evidence and current European League Against Rheumatism (EULAR) recommendations, azathioprine is recommended as first-line preventive therapy for both parenchymal and vascular involvement in NBD [11, 15–19].

The recommended starting dose of azathioprine is 50 mg/day in two divided doses. This dose can be escalated stepwise to a maintenance dose of 2.5 mg/kg/day in 1 month. Liver function tests (LFT) and complete blood counts (CBC) must be followed during the escalation period to prevent toxicity. In patients with genotypic variations of the enzyme thiopurine S-methyltransferase, which methylates azathioprine, severe bone marrow suppression and gastrointestinal side effects may occur that may preclude the use of azathioprine in those patients. The therapeutic effect of azathioprine can be monitored with the increase in mean corpuscular volume. In the case of leukopenia or thrombocytopenia, the dose of azathioprine should be modified. If severe leukopenia (WBC < 1000/μL) and thrombocytopenia (platelet count $<50 \times 10^9/\mu L$) occur, the drug should be discontinued. Additionally, idiosyncratic agranulocytosis can be seen in the first few weeks of therapy. Patients showing acute hypersensitivity reactions with severe nausea, vomiting, diarrhea, fever, rash, urticaria, malaise, myalgia, and arthralgia to azathioprine should be switched to another treatment alternative. Azathioprine may increase the risk of malignancies, such as lymphoma and skin cancer, particularly after 10 years of treatment. Additionally, there are other rare side effects of azathioprine, e.g., pancreatitis, reduced resistance to infection, and teratogenesis. Azathioprine is in category D for pregnancy.

Mycophenolate Mofetil

Mycophenolate mofetil (MMF) is a prodrug of mycophenolic acid, which is a selective, potent, but reversible inhibitor of the enzyme inosine monophosphate dehydrogenase (IMPDH). There are several mechanisms of action of MMF; however, the immunosuppressive activity is based on de novo synthesis of the guanosine nucleotides, and MMF inhibits mitogen and antigen-stimulated proliferation of B and T lymphocytes [20]. Mycophenolic acid is a potent inhibitor of the type II isoform of IMPDH, which is expressed in activated lymphocytes; therefore, it has a more potent cytostatic effect on lymphocytes than on other cell types. It also inhibits the synthesis of glycoproteins and the expression of some adhesion molecules, thereby decreasing the recruitment of lymphocytes and monocytes into sites of inflammation [21]. Besides, MMF inhibits the release of tumor necrosis factor-alpha and promotes the production of the anti-inflammatory interleukin-10 [22].

There have been no RCTs of MMF use in BD and NBD. Several reports have been published showing steroid-sparing effects in the treatment of uveitis and its use for pulmonary involvement of BD. There are only a few case reports on the administration of MMF in NBD, suggesting MMF is at least as effective as azathioprine and has a more favorable side effect profile than azathioprine [3, 23, 24]. According to these data, MMF is recommended to use as an alternative option in NBD (level of evidence: IIA, strength of recommendation: C) [25, 26].

The recommended starting dose of MMF is 1 gr/day in two divided doses. The dose can be increased to a maintenance dose of 2 gr/day in 4 weeks. MMF may cause myelosuppression, and long-term use may cause liver dysfunction; therefore, monthly LFT and CBC for 3 months and every 3 months thereafter is recommended. Other side effects of MMF are asthma, infections (respiratory tract, herpes simplex infections, and oral candidiasis), metabolic acidosis, and pleural effusion. MMF may also increase the incidence of skin cancers. These adverse effects, however, are more favorable compared to those of other immunosuppressants [23]. MMF is teratogenic, and therefore, it is in category D for pregnancy.

Methotrexate

Methotrexate (MTX) is a folate derivative that inhibits dihydrofolate reductase (DHFR). This inhibition results in subsequent depletion of tetrahydrofolate (THF) required for purine and pyrimidine synthesis. Therefore, MTX shows its anticancer effect by inhibiting cell growth and proliferation and also its anti-inflammatory effect by inhibition of lymphocyte multiplication. There has been no RCT on the use of MTX in BD and NBD. Administration of MTX has been advocated in ocular manifestations of BD in several reports [28–32] and a longitudinal study on posterior uveitis for up to 15 years [33]. MTX has been studied in a small number of patients with intestinal BD, suggesting that it may be an option in intolerant/refractory cases to azathioprine [34]. The first report on NBD showed low-dose weekly MTX therapy might have a beneficial effect in the treatment of progressive NBD [35]. Likewise, an open trial with a small number of patients showed that MTX might prevent the progression of chronic progressive NBD [8]. A retrospective study on NBD showed that among various immunosuppressive drugs, only MTX significantly improves the outcome of the patients by reducing mortality and disability [9]. According to these data, MTX is recommended to use as an alternative option in NBD (level of evidence: IIA, strength of recommendation: C) [26, 27].

The recommended administration of MTX is 7.5–25 mg once weekly as a single oral dose or in divided doses given over 24 hours. Methotrexate can be started with 7.5 mg/week and increased stepwise by 2.5 mg/week. The remission is usually achieved by a dose of 12.5–25 mg/week. Before starting the therapy, CBC, LFT, renal function tests (RFT), chest X-ray (CXR), hepatitis B and C evaluation, and infectious workup should be done. Significant side effects of MTX are bone marrow suppression, hepatotoxicity, opportunistic infections, and mucositis. CBC, LFT, and serum creatinine should be monitored every 2–4 weeks for the first 3 months, then every 3 months. Minor side effects are malaise, nausea, vomiting, diarrhea, headache, alopecia, fatigue, mood alteration, dizziness, fever, myalgias, and polyarthralgia, which are common but usually responsive to folate supplementation. Most minor toxic effects are associated with depletion of folate. Folate supplementation with 1 mg daily or 7 mg once weekly should be considered for all patients. MTX is in pregnancy category X.

Cyclophosphamide

Cyclophosphamide (CYC) is a fast-acting alkylating agent that induces apoptosis and decreases DNA synthesis. It is a potent immunosuppressive agent acting on rapidly dividing cells, including T and B lymphocytes. To date, there has been no RCT on its use in BD and NBD. However, several studies recommend the use of CYC in NBD. It has been found as a beneficial therapeutic agent in BD mainly for refractory retinal vasculitis and uveitis, and systemic vasculitis (arteritis, venous thrombosis-superior vena cava or Budd–Chiari syndrome, central or peripheral arterial aneurysm, and neurologic involvement) [36–41]. A retrospective study that includes 40 patients with NBD showed significant clinical improvement with CYC treatment [42]. A French study involving 115 NBD patients compared i.v. CYC (n = 53) with azathioprine (n = 40) and steroids alone (n = 19) and found a trend of longer event-free survival in patients receiving CYC in comparison to patients receiving azathioprine [43]. However, in a South Korean study on 22 p-NBD patients, treatment with CYC together with steroids was found to be as effective as treatment with steroids alone in preventing relapses [44]. There are also reports demonstrating benefits in p-NBD [3, 45]. CYC is recommended to use as a second-line treatment choice for NBD (level of evidence: IIA, strength of recommendation: C) [26]. CYC should be considered only in selected cases with clinically refractory disease due to its severe toxicity.

Conventionally, CYC is administered as a dose of 1000 mg/month i.v. for 6–12 months or 0.8 g/m^2 for the treatment of NBD. It is recommended to administer monthly i.v. CYC in combination with glucocorticoids for patients with severe CNS involvement. Major side effects of CYC are myelosuppression, pulmonary fibrosis, renal toxicity, cardiotoxicity, hemorrhagic cystitis, infertility, and malignancy (especially bladder cancer). The most common side effects are nausea, vomiting, diarrhea, loss of appetite, alopecia, amenorrhea, leukopenia, and hemorrhagic cystitis. During i.v. infusions, CYC is generally administered with antiemetics and mesna for the prevention of nausea and hemorrhagic cystitis, respectively [46]. Baseline infectious workup should be done, including hepatitis B and C and latent tuberculosis. For young patients planning to have a child, referral to a fertility clinic for the preservation of sperm or ova should be discussed with the patients. For follow-up, CBC and urine analysis before every cycle should be performed. Dose adjustments may be required to maintain a WBC count of >3500 cells/μl. Urine cytology every 6–12 months and cystoscopy of patients with hematuria or abnormal cytology should be implemented. CYC is in pregnancy category X.

Infliximab

Infliximab (IFX), a chimeric IgG1 monoclonal antibody that binds with high affinity to both soluble and transmembrane forms of TNF. It neutralizes TNF's biological activity, causing cytotoxicity by binding transmembrane TNF on monocytes and macrophages. IFX is useful in various manifestations of BD. According to current EULAR recommendations, IFX should be considered in selected cases for mucocutaneous lesions (level of evidence: IB, strength of recommendation: A), posterior segment eye involvement (level of evidence: IIA, strength of recommendation: A/B), initial or recurrent episodes of acute sight-threatening uveitis (level of evidence: IIA, strength of recommendation: B), refractory venous thrombosis, refractory arterial involvement, severe or refractory gastrointestinal involvement, and severe or refractory patients with NBD (level of evidence: III, strength of recommendation: C) [11].

There are two observational studies on mucocutaneous lesions [47, 48] and retrospective studies on extra-parenchymal neurologic and vascular involvement [49–52], showing its efficacy in BD. The efficacy of IFX has been demonstrated in treating BD-related ocular involvement (uveoretinitis and panuveitis) in a large number of case reports, open-label clinical trials, and non-randomized comparative studies [47, 53–57]. Most of the patients with uveitis treated with IFX showed complete response and improvement in visual acuity. In two recent studies, it has been reported that earlier initiation of IFX in uveitis led to a better visual outcome [58, 59]. Recent retrospective studies substantiate the beneficial effect of IFX in patients with refractory gastrointestinal involvement [60–63]. In line with these findings, 20–61% complete remission rates were observed in patients with refractory gastrointestinal disease who received anti-TNF-α treatments [64, 65].

There are 15 case reports [65–79], four case series [80–84], two observational studies [85, 86], and a phase III trial (included only three NBD patients out of 18 with various manifestations) [87] demonstrating the beneficial effects of IFX therapy mainly in parenchymal NBD. In these studies, a total of approximately 110 NBD patients were reported, and most of them responded well to first- or second-line IFX treatment in terms of clinical and radiological remission rate. Infliximab also showed a corticosteroid-sparing effect and a rapid onset of action. Similarly, three studies demonstrated long-term remission on IFX treatment for up to 4 years [74, 84, 85]. Zeydan et al. reported that other than 15 NBD patients, 74 patients with BD without NBD also received IFX within the same study period. At the time of the last follow-up, none of these patients had developed NBD. According to the authors' observation, early use of infliximab can be beneficial in the prevention of NBD. No significant side effects were seen other than one patient who developed pulmonary and CNS tuberculosis [84].

The optimal time for the duration of IFX treatment is still unclear. One study reported that the drug-free long-term remission was achieved at the end of the second year [88]. IFX is effective and well-tolerated in all major organ involvements of BD. Infliximab is currently recommended for patients refractory to conventional treatments or as first-line therapy in patients with severe disease and poor prognostic factors. It improves clinical and immunological parameters and ameliorates the level of disability when initiated early.

IFX is given at a dose of 5 mg/kg by i.v. infusions at 0, 2nd, and 6th weeks and then repeated every 8 weeks. The most common side effects of IFX are infusion site reactions, elevated liver enzymes, leukopenia, hypertension, hyperglycemia, and fever. Major side effects are infections (varicella-zoster infection, reactivation of tuberculosis, and other opportunistic infections), lymphoma, osteomalacia, and CNS demyelinating disease. Before the administration of IFX, baseline infectious workup and, due to the risk of tuberculosis reactivation and hepatitis, screening tests (hepatitis B and C, PPD, or QuantiFERON-TB test) are necessary. Isoniazid prophylaxis (300 mg/day) should be prescribed for 6–9 months in patients with latent tuberculosis [89]. Live vaccines are contraindicated during the treatment. IFX has been categorized as category B for pregnancy.

Etanercept

Etanercept (ETN) acts as a soluble fusion protein that blocks the interaction of TNF with cell surface TNF receptors p55 and p75. Etanercept is an inhibitor of TNF, however not an antibody to the molecule that renders TNF biologically inactive by preventing binding to both TNF receptors. Etanercept is the only TNF-α agent that has an RCT in BD. A double-blind RCT of etanercept (25 mg twice/week, for 4 weeks) in 40 BD patients with various manifestations (pathergy, mucocutaneous lesions, and arthritis) was effective in decreasing the frequency of oral ulcers and nodular and papulopustular lesions. The drug decreased the number of genital ulcers and arthritis episodes during the treatment period; however, the difference was not statistically significant. Also, no significant changes in the pathergy response could be found. The reason for inefficacy on genital ulcers and arthritis could be due to the small sample size, short duration of the trial, or the high placebo response rate. In this trial, one patient was withdrawn due to infectious colitis, and another patient developed gastrointestinal involvement. Interestingly, some patients had worsening symptoms 3 months after ETN was discontinued [90]. Additionally, there are many case reports of ETN showing a rapid resolution of severe mucocutaneous lesions in BD [47, 91–94]. ETN may also have beneficial effects on refractory arthritis [92] and ocular manifestations [95]. In an open retrospective study, higher clinical remission rate and healing of intestinal ulcers were observed with ETN treatment (25 mg twice a week) in comparison to conventional treatment (MTX 15 mg/week or prednisolone) without any withdrawals due to toxicity [96]. In NBD, a case report suggests that ETN is beneficial in treating refractory parenchymal neurological involvement [97]. Hence, a recommendation was made for ETN as an alternative treatment option in refractory NBD [98]. There are also several case reports in pediatric patients reporting favorable results with ETN without any serious adverse events [99, 100].

ETN is administered 25 mg subcutaneously (s.c.) twice weekly or 50 mg s.c. once weekly. The most common side effects are injection site reactions, elevated liver enzymes, leukopenia, hypertension, hyperglycemia, and fever. Major side effects are infections (varicella-zoster infection, reactivation of tuberculosis, and other opportunistic infections), lymphoma, osteomalacia, and demyelinating CNS disease. Before the administration of ETN, baseline infectious workup and, due to the risk of activating latent tuberculosis and hepatitis, screening tests (hepatitis B and C, PPD, or QuantiFERON-TB test) should be performed. Isoniazid prophylaxis (300 mg/day) should be prescribed for 6–9 months in patients with latent tuberculosis. Live vaccines are contraindicated during the treatment. ETN has been categorized as category B for pregnancy.

Adalimumab

Adalimumab (ADA) is a recombinant human IgG1 monoclonal antibody specific for human TNF-α. It binds to TNF-α and blocks its interaction with the p55 and p75 cell surface TNF receptors, resulting in lysis of cells with surface TNF in the presence of complement. In two phase III RCTs, ADA was effective in the treatment of non-infectious, non-anterior uveitis in a small subset of BD patients [101, 102]. Although the study had limitations such as a limited number of patients and the data on BD patients was not reported separately, the growing observational evidence confirmed the efficacy of ADA in BD uveitis [103–105]. Based on these studies, ADA has been approved by the European Medicines Evaluation Agency (EMEA) and the US Food and Drug Administration (FDA) for the treatment of non-infectious intermediate, posterior, and panuveitis. In a recent observational study including 106 patients with uveitis, ADA was associated with amelioration of inflammation along with a corticosteroid-sparing effect and good preservation of visual acuity even in the absence of concomitant other immunosuppressive treatments [106].

Additionally, two recently published multicenter observational studies [107, 108] also supported the effectiveness of IFX and ADA in the treatment of BD uveitis. Both have significantly improved the outcome of BD patients refractory to classical immunosuppressives. Not surprisingly, some patients were unresponsive to IFX and ADA.

A retrospective study from a single center assessed ADA-based regimens (either single or combined with immunosuppressive drugs) compared with immunosuppressive therapies (azathioprine, cyclosporine A, CYC, and MTX) and the corticosteroid-sparing effect of ADA in 70 patients with recurrent venous thrombosis in BD [109]. ADA-based regimens provided a better rapid vascular response compared to other immunosuppressive therapies during a mean follow-up of 26 months, allowing less exposure to corticosteroids. No differences in efficacy were demonstrated regarding anticoagulant use, but there was a lower number of patients requiring additional anticoagulants. Another multicenter, retrospective study assessed the efficacy of IFX and ADA (either single or combined with immunosuppressives and corticosteroids) in 18 BD patients (IFX = 15, ADA = 3) with refractory vascular involvement including arterial aneurysms, arterial occlusions, intracardiac thrombosis, and large vein thrombosis [86]. The vascular response was achieved in 16 patients allowing less corticosteroid dependency during a mean follow-up of 15 months. According to these results, ADA seems to be more effective for venous thrombosis than classical immunosuppressives. Other than ocular and vascular involvement, in an open-label study, the efficacy of ADA was demonstrated in 20 patients with intestinal BD refractory to corticosteroids and standard immunosuppressive therapies. Complete remission was achieved in 20% of the patients, and 61% of the patients were able to discontinue steroids in 1 year [110]. There are further studies on the use of ADA in the treatment of intestinal BD [111, 112]. Several case reports demonstrated that ADA might also be beneficial for recalcitrant leg and genital ulcers, cerebral vasculitis, and pulmonary artery aneurysm in patients with BD [113–117].

In NBD, the use of ADA is based on case reports and observational studies. In the observational, multicenter study, 17 BD patients with severe and refractory parenchymal NBD were treated with an anti-TNF-α (IFX, $n = 13$; ADA, $n = 4$). Complete clinical and radiological remission at 12 months was achieved in six patients, and partial response was observed in ten patients. There are also case reports and clinical studies with favorable results about the use of ADA in pediatric patients [118–123]. TNF blockade seems to act as an effective therapeutic approach for patients with severe onset or refractory NBD, a difficult to treat population [52]. ADA can be used alone or in combination with immunosuppressive drugs, which might be more effective. ADA seems to be a feasible option for NBD patients who are refractory or intolerant to infliximab [124].

ADA is administered at a dose of 40 mg s.c. every 2 weeks. The most common side effects are injection site reactions, elevated liver enzymes, leukopenia, hypertension, hyperglycemia, and fever. Major side effects are infections (varicella-zoster infection, reactivation of tuberculosis, and other opportunistic infections), lymphoma, osteomalacia, and CNS demyelinating disease. Before the administration of ADA, baseline infectious workup and, due to the risk of reactivation of latent tuberculosis and hepatitis, screening tests (hepatitis B and C, PPD, or QuantiFERON-TB test) should be performed. Isoniazid prophylaxis (300 mg/day) should be prescribed for 6–9 months in patients with latent tuberculosis. Live vaccines should not be used during treatment. ADA has been categorized as category B for pregnancy.

Anti-IL1 Therapies

IL-1 is one of the major cytokines in the pathogenesis of BD. IL-1α and IL-1β are cytokines of the IL-1 family with prominent proinflammatory effects. IL-1α and IL-1β bind to the same receptor molecule, which is called type I IL-1 receptor (IL-1RI). IL-1α or IL-1β bind to the first extracellular chain of IL-1RI that recruits the IL-1 receptor accessory protein (IL-1RAcP), which serves as a coreceptor for signal transduction. Recently, numerous IL-1 blockers were developed, and anakinra (ANA), canakinumab (CAN), and gevokizumab (GEV) have been tried in BD.

Anakinra

Anakinra (ANA) is the recombinant form of the natural IL-1 receptor antagonist (IL-1Ra) which acts by preventing the binding of IL-1α and IL-1β to IL-1R1. To date, there has been no RCT on its use in BD or NBD. However, there are several case reports or case series, one observational cohort study, and a clinical trial on the use of ANA in BD. In 2008, the first case report demonstrated that ANA swiftly controlled clinical and inflammatory biomarkers in a patient with refractory mucocutaneous BD [125].

In 2015, the first case series of nine refractory BD patients reported the use of ANA in patients with active mucocutaneous lesions in addition to major organ involvements. Eight of those patients improved promptly within 2 weeks; however, all experienced at least one relapse during a mean follow-up of 29 weeks. Therefore, four patients were switched to another agent eventually (one CAN, one ADA, and two CYC) [126]. Another multicenter, retrospective study evaluated the efficacy and safety of IL-1 antagonists. Treatment discontinuation during ANA or CAN therapy was reported in seven out of 18 subjects [127]. A retrospective cohort study of 30 BD patients with varied manifestations (mostly ocular involvement, two patients with NBD) were treated with either ANA 100 mg/day ($n = 27$) or CAN 150 mg every 6–8 weeks ($n = 3$). Both drugs showed a favorable efficacy (13/30 complete remission) and safety that is supported by an overall cumulative survival of 67.8% at 24 months [128]. A small prospective open-label study evaluated the safety, efficacy, and dose of ANA for the treatment of refractory oral and genital ulcers in six patients with BD. The primary outcome was defined as no ulcers for two consecutive months. The dose of ANA had to be increased from 100 mg to 200 mg/day due to inadequate response during the first month in all patients. The study demonstrated partial efficacy on resistant oral and genital ulcers (complete response = 2, partial response = 2, treatment failures = 2) [129].

Another multicenter, retrospective, observational study assessed the efficacy of ANA and CAN in 19 BD patients with uveitis. Thirteen patients received ANA, and six received CAN. In 37% of the patients, IL-1 inhibition was used as the first-line biological treatment. It is reported that ocular inflammatory flares and frequency of retinal vasculitis significantly decreased at month 12 [130]. Interestingly, the use of IL-1 inhibitors in combination with any immunosuppressive agent did not add benefit when compared to IL-1 inhibitors alone. In a retrospective observational study, the use of ANA or CAN in patients with refractory uveitis was associated with a complete response in 25 out of 36 patients. Surprisingly, patients with longer disease duration had better responses. Steroid dosage was significantly decreased at 12 months compared to baseline. ANA was discontinued only in three out of 13 patients due to inefficacy. CAN was stopped in one out of six subjects because of an attack in the CNS [131]. In a case report, a patient with BD-related sacroiliitis who is refractory to infliximab and concomitant prednisone (50 mg/day) showed complete remission with ANA within a few days (100 mg/day). The prednisone dose was tapered to 5 mg/day without any relapses [132].

To date, there is only one report on the use of ANA in patients with NBD. In this report, ANA (100 mg/day) was administered to a 30-year-old female patient with NBD and refractory uveitis. Rapid response was seen in clinical and radiological findings within a few days, allowing steroid tapering. However, the patient relapsed with bilateral uveitis after 18 months of treatment, and ANA was switched to CAN (150 mg every 4 weeks). Later, CAN was also discontinued because of poor control of the neurological involvement [133]. According to this data, ANA might have a potential benefit for ocular, mucocutaneous, and articular manifestations of BD. However, unresponsiveness to ANA has also been reported, but increasing the dosage to more than 100 mg/day or switching to another anti-IL-1 agent could be an option to overcome inefficacy. Also, since IL-1 inhibition does not increase the risk of tuberculosis reactivation, IL-1 inhibition can be considered as a treatment option where both BD and tuberculosis are prevalent.

ANA is used at a dose of 100 mg/day or 2 mg/kg/day by s.c. injections. The dose can be increased to 200 mg/day in case of partial clinical response. The most common adverse event is injection site reactions. Additionally, ANA was associated with diffuse pruritic urticarial lesions. Although ANA is relatively safe compared to traditional drugs used in NBD, opportunistic respiratory tract infections are not rare.

Canakinumab

Canakinumab (CAN) is a highly specific, fully human IL-1b monoclonal antibody. It selectively neutralizes the bioactivity of human IL-1b, inhibiting its binding to IL-1RI and IL-1RII. In a case report, a single dose of 150 mg of CAN was found effective in a 16-year-old patient with Behçet's uveitis refractory to TNF antagonists (IFX, ADA) and ANA in inducing clinical manifestations, even the ocular symptoms, and in all inflammation markers within a few weeks. However, the follow-up period was only 8 weeks [134]. In another case report, CAN 150 mg/8 weeks was prescribed after the failure of IFX and ADA treatment and accounted for reasonable control of ocular symptoms and the absence of flares in the first 6 months of treatment [135].

In 2014, three additional patients with mucocutaneous and other various involvements were reported with complete remission during a 6–12-month follow-up. However, deep vein thrombosis occurred in a patient who was on CAN treatment every 8 weeks. Interestingly, remission of that patient was achieved when CAN was administered every 6 weeks [136]. CAN (4 mg/kg every 28 days) was effectively used in a 9-year-old boy with mucocutaneous, ocular findings and fever who failed previous ANA treatment. Complete clinical and laboratory remission was obtained in 4 weeks, and steroid was gradually tapered; the patient remained utterly asymptomatic during a 6-month follow-up [137]. A multicenter, retrospective study evaluating the efficacy of both ANA and CAN in a cohort of 30 BD patients showed that BD patients with an initial low response to ANA recovered completely following an increase in the drug dose. Also, complete response was seen by switching to CAN after ANA failure and adjusting the dose of CAN from 8 weeks to 6 weeks in patients experiencing a disease relapse while on canakinumab. The median time of response to therapy was 6 weeks for ANA and 3 weeks for CAN. Besides, treatment survival was acceptable for both IL-1 inhibitors evaluated, and no adverse events were recorded [138]. In a report, CAN successfully controlled ocular manifestations in a patient discontinuing ANA for adverse events, with complete ocular control at 36-month follow-up [139]. In another report aforementioned in the ANA section, CAN (150 mg every 4 weeks) was given to a 30-year-old woman with NBD also with refractory ocular involvement, following ANA ocular failure. In this patient, CAN controlled ocular findings; however, it was discontinued because of poor control of the neurologic involvement [136]. According to these results, CAN has shown a potential role in the management of BD. In some patients, the administration of 150 mg every 6 weeks instead of 8 weeks led to better results.

The starting dose of CAN is 300 mg i.v. The initial dose can be continued as 150 mg i.v. infusions every 4 weeks for 6 months. Depending on the response in the first 6-month period, 150 mg CAN can be administered as s.c. injections or i.v. infusions and every 4 weeks thereafter. CAN has an excellent safety profile with no serious adverse effects. It offers a considerable advantage over ANA, which must be injected daily and which is often poorly tolerated by patients.

Anti-IL-6

Interleukin-6 (IL-6) is a proinflammatory cytokine secreted by T and B cells, monocytes, macrophages, and synovial fibroblasts. It mediates a variety of immunologic responses. IL-6 has a critical role in the differentiation of CD4-positive T helper cells into Th17 cells, antibody production by B cells, and differentiation of macrophages, and it also inhibits the differentiation of regulatory T cells. Also, IL-6 increases vascular permeability and angiogenesis. The IL-6 receptor has a receptor and a signal transducer component. The receptor component has transmembrane and soluble forms. Interaction between two IL-6 molecules with two IL-6 receptor and two signal transducer components induces intracellular signal processing. Inhibition of IL-6 signaling reduces inflammation by reducing autoreactive Th17 and Th1 cells and suppressing autoantibody production. Additionally, anti-IL-6 therapies reduce hepatic acute-phase proteins and have anti-angiogenic effects.

IL-6 pathway has a critical role in the pathogenesis of NBD. The serum level of IL-6 is elevated in patients with active BD, and it has been shown that IL-6 is the major cytokine to be increased in NBD during active disease [140]. Therefore, it is rational to target the IL-6 pathway as a therapeutic regimen in the treatment of NBD.

Tocilizumab

Tocilizumab (TCZ) is a fully humanized antibody against soluble and membrane-bound IL-6 receptors. There has been no RCT of its use in BD and NBD. Hence, the clinical evidence comes mainly from several case reports and an observational study regarding the use of TCZ in BD. There are several case reports and a small case series on the use of TCZ in patients with refractory uveitis to interferon-alpha (IFN-α) and IFX, suggesting that this agent may be useful in reducing relapses, protecting visual acuity, reducing central macular thickness, and improving angiographic lesions [141–146]. In another multicenter, retrospective study, 11 patients with severe refractory uveitis were reported. Interestingly, all ocular outcome measures in all patients responded rapidly after TCZ [147]. Besides, extraocular manifestations were improved in three patients, but a worsening of arthritis was seen in one patient. Likewise, worsening of mucocutaneous manifestations under treatment with TCZ has also been reported [148–150]. Despite the variability of effects of TCZ on different organ involvements of BD, TCZ seems to be an option for severe refractory uveitis of BD.

A small retrospective study reported the efficacy of TCZ in seven refractory vascular-BD patients with multiple arterial (aneurysms, occlusions, stenosis) and venous lesions [151]. In this study, TCZ was used as add-on therapy. After a mean follow-up of 19 months, all patients showed improvement in clinical symptoms and inflammatory biomarkers. No new onset arterial or venous lesions were observed during the follow-up of the patients, and all but one showed either complete or partial response. Treatment with TCZ also allowed a reduction in doses of previous therapies, including corticosteroids. These findings, from this small, open-label, uncontrolled series, suggest that TCZ might be an alternative to anti-TNF treatments for BD with refractory vascular involvement.

In NBD pathogenesis, CSF IL-6 level is reported as a possible biomarker of disease activity and long-term outcome [152, 153]. Therefore, TCZ therapy seems to be a promising option for CNS involvement. However, there are only case reports of refractory NBD that were successfully treated with TCZ [154–156]. According to these reports, TCZ was well tolerated without serious adverse events. An increasing number of studies suggested TCZ as a feasible therapeutic opportunity in BD. TCZ seems to be beneficial mainly for ocular involvement and refractory NBD. Therefore, TCZ is currently recommended to use as an alternative treatment choice for NBD therapy in clinical practice (level of evidence: IIA, strength of recommendation: C) [26]. Also, TCZ might be an alternative to anti-TNF agents for patients with arterial involvement refractory to immunosuppressives [157]. However, caution is advised for patients with skin and mucosal manifestations since TCZ may even worsen the symptoms in those cases.

TCZ is administrated as 4–8 mg/kg i.v. infusion every 4 weeks. Before starting the therapy, LFT, CBC, lipid panel, hepatitis B and C, PPD, and QuantiFERON-TB tests should be performed. Usually, influenza and pneumococcal vaccinations are also recommended. TCZ treatment may exhibit various adverse effects such as infections, gastrointestinal alterations, increased hepatic transaminase levels, hepatotoxicity, abnormalities in lipid panel, neutropenia, thrombocytopenia, gastrointestinal perforation, and infusion-related hypersensitivity reactions. On follow-up, CBC, hepatic function tests at 4–8 weeks and then every 3 months, and lipid panel at 4–8 weeks and then every 6 months should be monitored.

Interferon-α

Interferons are naturally occurring glycoproteins that have antiviral, antitumor, and immunomodulatory actions. Both interferon (IFN)-α-2a and -α-2b are the two forms being used in BD; however, IFN-α-2a seems to be more effective. Rapid mechanism of action and long-term remission after treatment cessation are the most remarkable characteristics of IFNs. IFN-α has been in use since 1986, especially for the treatment of BD-related uveitis affecting the posterior segment. There is an RCT [158] and an observational study [159] showing efficacy for mucocutaneous lesions. There are also studies showing beneficial effects on mucocutaneous, articular, and ocular manifestations of BD [36, 160, 161]. In a comparative study, IFX ($n = 20$) was compared with IFN-α-2a ($n = 33$) in patients with refractory BD uveitis [162]. According to the results of the study, both agents were found effective, and no major difference between IFX and IFN in controlling intraocular inflammation was noted. Although the experience with IFN in NBD is limited to case reports or series, it is an available choice in p-NBD, and it may only be considered as an option for second-line treatments in patients who cannot receive anti-TNF treatments ([16, 161, 163, 164]). It is also reported that INF can be used even in refractory NBD in combination with cyclophosphamide (CYC) [165, 166]. A retrospective study investigated the efficacy and safety of add-on IFN (together with corticosteroids, cyclosporine, AZA, CYC, MTX, or tacrolimus) in 30 BD patients with uveitis refractory to steroids [167]. This study did not report any safety issues in addition to the beneficial effects. In a retrospective study, 36 patients with BD-related uveitis were treated with IFN. Of 36 patients, 31 responded to therapy, and of the 21 patients who discontinued IFN therapy, 76% did not have a relapse within 5 years of follow-up following discontinuation.

According to the current EULAR recommendations, IFN should be considered in selected cases for mucocutaneous lesions (level of evidence: IB, strength of recommendation: A). Patients presenting with an initial or recurrent episode of acute sight-threatening uveitis should be treated with IFN (level of evidence: IIA, strength of recommendation: B) [11]. Also, it has been stated that treatment with IFN-α significantly reduced the duration and pain of oral and genital ulcers and reduced the severity and rate of recurrence of attacks of the eye manifestations.

IFN-α is commonly used as 6–9 MIU/day s.c. for 1 week and tapered to 3 MIU s.c. three times a week, according to the patient's response. The side effects of IFN are flu-like symptoms, injection site reactions, mild leukopenia, asymptomatic elevated liver enzymes, alopecia, depression, and thyroid dysfunction. IFN generally has a favorable side effect profile. Nevertheless, tolerability issues can limit its use [168, 169].

Hematopoietic Stem Cell Transplantation

In NBD, hematopoietic stem cell transplantation (HSCT) has been proposed as a therapeutic tool for cases with severe and refractory disease course. Currently, there have been no RCTs on HSCT in patients with BD. Nevertheless, there is a growing body of evidence on the effect of HSCT in BD patients with either refractory disease course or coexisting hematological diseases. Since 2001, there have been 20 studies in which 29 HSCT procedures (14 autologous, nine allogeneic, three cord blood, three not defined) were performed in 27 BD patients (14 female and 13 male) [170–186]. Among these 27 patients, two patients had undergone first an autologous HSCT and then allogeneic stem cell transplantation. The mean follow-up period of the patients after transplantation was 36 months (range 2–78). The indication for transplantation was refractory BD in 13 patients and accompanying hematological disorders (myelodysplastic syndrome and acute myeloid leukemia) in 14 patients.

Of 13 patients who received 15 HSCT procedures (13 autologous and two allogeneic), eight were transplanted for NBD, four for pulmonary artery aneurysms and concomitant intracardiac thrombosis, and one for gastrointestinal involvement. Of 20 patients who had hematological disorders, ten patients had refractory BD, nine had BD-related gastrointestinal involvement, and one had BD-related uveitis. Of eight patients who were transplanted for refractory NBD, four had complete remission. Two patients remained refractory to autologous HSCT and underwent allogeneic stem cell transplantation, but both patients relapsed after approximately 2 years. One did not achieve full remission after autologous HSCT. Two patients with refractory pulmonary artery aneurysms had complete remission of aneurysms. However, one of the patients continued to have mucocutaneous lesions after transplantation. Infections were the most common cause of transplant-related mortality. Significant complications of allogeneic HSCT are infections, graft-versus-host disease (GVHD), and hepatic, renal, and pulmonary damage. The patients with two transplantations experienced no complications during autologous HSCT, while grade 3 and 4 gastrointestinal and liver GVHD occurred after allogeneic HSCT. One patient died due to an opportunistic infection [177]. Among the 27 patients, three patients developed mild gastrointestinal side effects, and four had a neutropenic fever. One patient who underwent allogeneic HSCT died of an infection after 2 months.

In conclusion, HSCT may be a therapeutic option for patients with severe or refractory BD. Usually, autologous HSCT has been performed because it is relatively safer than allogeneic HSCT. Although HSCT is an effective treatment modality, it should be performed only in selected cases weighing the possible risks and benefits which can differ among transplant centers.

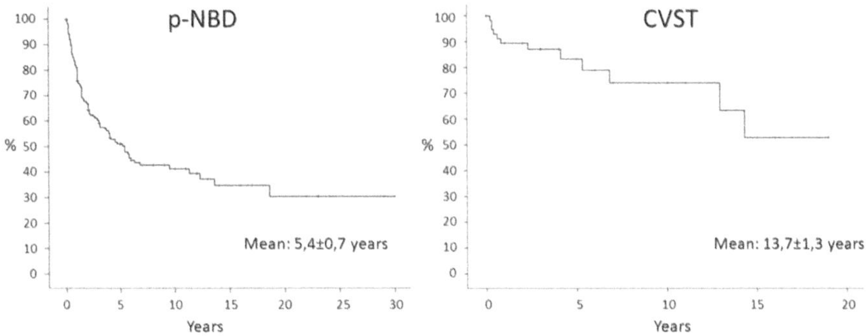

Fig. 1 Relapse-free survival of patients with parenchymal neuro-Behçet's disease (p-NBD, on the left) and cerebral venous sinus thrombosis (CVST, on the right)

How Long Should We Treat?

Of 430 patients followed at Istanbul University, at least one relapse was observed in 42% of p-NBD patients and 14% of CVST patients after a mean follow-up period of 5 years [187]. The number of relapses was more than two in 42% of relapsing patients even under treatment. The mean time to relapse was approximately 5 years in the p-NBD group and 14 years in the CVST group. Additionally, the decrease in the annual attack risk below 5% occurred in the p-NBD and CVST group after the second and fifth years and below 2% in the seventh and eighth years, respectively (Fig. 1). Therefore, given the severe side effects of the treatments used in NBD and the high rate of disability after relapse, it is advisable to continue the treatments for at least 7 years in patients with p-NBD. For CSVT, on the other hand, it is plausible to stop the treatment after 2 years, because of the low rate of disability even after the second clinical attack. However, this decision must be done taking the presence of other vascular complications of BD into consideration.

References

1. Kotake S, Higashi K, Yoshikawa K, Sasamoto Y, Okamoto T, Matsuda H. Central nervous system symptoms in patients with Behçet disease receiving cyclosporine therapy. Ophthalmology. 1999;106(3):586–9.
2. Akman-Demir G, Ayranci O, Kurtuncu M, Vanli EN, Mutlu M, Tugal-Tutkun I. Cyclosporine for Behçet's uveitis: is it associated with an increased risk of neurological involvement? Clin Exp Rheumatol. 2008;26(4 Suppl 50):S84–90.
3. Kürtüncü M, Tüzün E, Akman-Demir G. Behçet's disease and nervous system involvement. Curr Treat Options Neurol. 2016;18(5):19.
4. Sorensen PS, Mellgren SI, Svenningsson A, Elovaara I, Frederiksen JL, Beiske AG, Myhr KM, Søgaard LV, Olsen IC, Sandberg-Wollheim M. NORdic trial of oral Methylprednisolone as add-on therapy to Interferon beta-1a for treatment of relapsing-remitting Multiple Sclerosis (NORMIMS study): a randomized, placebo-controlled trial. Lancet Neurol. 2009;8(6):519–29.

5. Seider N, Beiran I, Scharf J, Miller B. Intravenous immunoglobulin therapy for resistant ocular Behçet's disease. Br J Ophthalmol. 2001;85(11):1287–8.
6. Cantarini L, Stromillo ML, Vitale A, Lopalco G, Emmi G, Silvestri E, Federico A, Galeazzi M, Iannone F, De Stefano N. Efficacy and safety of intravenous immunoglobulin treatment in refractory Behcet's disease with different organ involvement: a case series. Isr Med Assoc J. 2016;18(3–4):238–42.
7. Noel N, Bernard R, Wechsler B, Resche-Rigon M, Depaz R, Le Thi Huong Boutin D, et al. Long-term outcome of neuro-Behçet's disease. Arthritis Rheumatol. 2014;66D5:1306–14.
8. Kikuchi H, Aramaki K, Hirohata S. Low dose MTX for progressive neuro-Behçet's disease. A follow-up study for 4 years. Adv Exp Med Biol. 2003;528:575–8.
9. Hirohata S, Kikuchi H, Sawada T, Nagafuchi H, Kuwana M, Takeno M, et al. Retrospective analysis of long- term outcome of chronic progressive neurological manifestations in Behcet's disease. J Neurol Sci. 2015;349(1–2):143–8.
10. Uygunoğlu U, Siva A. Nervous system involvement in Behçet's syndrome. Curr Opin Rheumatol. 2019;31(1):32–9.
11. Hatemi G, Christensen R, Bang D, Bodaghi B, Celik AF, Fortune F, et al. 2018 update of the EULAR recommendations for the management of Behçet's syndrome. Ann Rheum Dis. 2018;77:808–18.
12. Yazici H, Pazarli H, Barnes CG, Tüzün Y, Özyazgan Y, Silman A, et al. A controlled trial of azathioprine in Behçet's syndrome. N Engl J Med. 1990;322:281–5.
13. Hamuryudan V, Ozyazgan Y, Hizli N, Mat C, Yurdakul S, Tüzün Y, et al. Azathioprine in Behcet's syndrome: effects on long-term prognosis. Arthritis Rheum. 1997;40:769–74.
14. Kurtuncu M, Tuzun E, Pehlivan M, et al. Clinical patterns and treatment effect on neuro-Behçet's disease: analysis of 351 patients. Clin Exp Rheumatol. 2008;26(Suppl 50):S43.
15. Yazici H, Seyahi E, Hatemi G, Yazici Y. Behçet syndrome: a contemporary view. Nat Rev Rheumatol. 2018;14:107–19.
16. Akman-Demir G, Saip S, Siva A. Behcet's disease. Curr Treat Options Neurol. 2011;13(3):290–310.
17. Hamuryudan V, Ozyazgan Y, Fresko Y, Mat C, Yurdakul S, Yazici H. Interferon alfa combined with azathioprine for the uveitis of Behçet's disease: an open study. Isr Med Assoc J. 2002;4:928–30.
18. Shahram F, Davatchi F, Chams H, Nadji A, Jamshidi A, Akbarian M, et al. Azathioprine and low dose pulse cyclophosphamide in severe ocular lesions of Behçet's disease. A preliminary report. Adv Exp Med Biol. 2003;528:571–3.
19. Saip S, Akman-Demir G, Siva A. Neuro-Behçet syndrome. Handb Clin Neurol. 2014;121:1703–23.
20. Allison AC, Almquist SJ, Muller CD, Eugui EM. In vitro immunosuppressive effects of mycophenolic acid and an ester prodrug, RS- 61443. Transplant Proc. 1991;23(suppl 2):10–4.
21. Allison AC, Eugui EM. Mycophenolate mofetil, a rationally designed immunosuppressive drug. Clin Transpl. 1993;7:96–112.
22. Durez P, Appelboom T, Pira C, Stordeur P, Vray B, Goldman M. Anti-inflammatory properties of mycophenolate mofetil in murine endotoxemia: inhibition of TNF-alpha and upregulation of IL-10 release. Int J Immunopharmacol. 1999;21:581–7.
23. Shugaiv E, Tüzün E, Mutlu M, Kiyat-Atamer A, Kurtuncu M, Akman-Demir G. Mycophenolate mofetil as a novel immunosuppressant in the treatment of neuro-Behçet's disease with parenchymal involvement: presentation of four cases. Clin Exp Rheumatol. 2011;29(4 Suppl 67):S64–7.
24. Ferreira BFA, Rodriguez EEC, Prado LLD, Gonçalves CR, Hirata CE, Yamamoto JH. Frosted branch angiitis and cerebral venous sinus thrombosis as an initial onset of neuro-Behçet's disease: a case report and review of the literature. J Med Case Rep. 2017;11(1):104.
25. Kalra S, Silman A, Akman-Demir G, Bohlega S, Borhani-Haghighi A, Constantinescu CS, et al. Diagnosis and management of Neuro-Behcet's disease: international consensus recommendations. J Neurol. 2014;261:1662–76.

26. Borhani Haghighi A, Kardeh B, Banerjee S, Yadollahikhales G, Safari A, Sahraian MA, Shapiro L. Neuro-Behçet's disease: an update on diagnosis, differential diagnosis, and treatment. Mult Scler Relat Disord. 2019;39:101906.
27. Shahram F, Davatchi F, Chams H, Akbarian M, Chams C, Tebbi ME. Methotrexate in ocular Behçet's disease, Preliminary report, Abstract FP-23. First APLAR Symposium on the Therapy of Rheumatic Diseases, Seoul, Korea, 1990.
28. Shahram F, Davatchi F, Chams H, Akbarian M, Chams C. Pulse cyclophosphamide versus pulse methotrexate in ophthalmological manifestations of Behcet's disease. In: Nasution AR, Darmawan J, Isbagio H, editors. APLAR rheumatology. Tokyo: Churchill Livingstone; 1992. p. 119–22.
29. Shahram F, Davatchi F, Chams H, Akbarian M, Gharibdoost F. Comparing 3 methods of Cytotoxic therapy in ophthalmologic lesions of Behcet's disease. In: Godeau P, Wechsler B, editors. Behcet's disease. Amsterdam: Elsevier Science Publishers BV; 1993. p. 635–40.
30. Shahram F, Chams H, Davatchi F, Nadji A, Jamshidi AR, Akbarian M. Intermediate-term results of combination therapy with low dose pulse cyclophosphamide and methotrexate in ocular lesions of Behcet's disease. In: Bang D, Lee ES, Lee S, editors. Behcet's disease. Seoul: Design Mecca Publishing; 2000. p. 453–6.
31. Davatchi F, Shahram F, Chams H, et al. Methotrexate for ocular lesions of Behçet's disease. Cohort study on 262 patients. Arthritis Rheum. 1998;41:S356.
32. Davatchi F, Shahram F, Chams H, et al. High dose methotrexate for ocular lesions of Behçet's disease. Preliminary short-term results. Adv Exp Med Biol. 2003;528:579–84.
33. Davatchi F, Shams H, Shahram F, et al. Methotrexate in ocular manifestations of Behçet's disease: a longitudinal study up to 15 years. Int J Rheum Dis. 2013;16:568–77.
34. Park J, Cheon JH, Park Y, Park SJ, Kim TI, Kim WH. Efficacy and tolerability of methotrexate therapy for refractory intestinal Behçet's disease: a single-center experience. Intest Res. 2018;16(2):315–8.
35. Hirohata S, Suda H, Hashimoto T. Low-dose weekly methotrexate for progressive neuropsychiatric manifestations in Behçet's disease. J Neurol Sci. 1998;159:181–5.
36. Saleh Z, Arayssi T. Update on the therapy of Behçet disease. Ther Adv Chronic Dis. 2014;5:112–34.
37. Hatemi G, Silman A, Bang D, et al. EULAR recommendations for the management of Behçet disease. Ann Rheum Dis. 2008;67:1656–62.
38. Hamuryudan V, Yurdakul S, Moral F, et al. Pulmonary arterial aneurysms in Behçet's syndrome: a report of 24 cases. Br J Rheumatol. 1994;33:48–51.
39. Hamuryudan V, et al. Pulmonary artery aneurysms in Behçet syndrome. Am J Med. 2004;117:867–70.
40. Davatchi F, Shahram F, Chams H, Akbarian M, et al. Pulse cyclophosphamide (PCP) for ocular lesions of Behçet's disease: double-blind crossover study. Arthritis Rheum. 1999;42:S320.
41. Davatchi F, Sadeghi Abdollahi B, Shams H, Shahram F, Nadji A, et al. Combination of pulse cyclophosphamide and azathioprine in ocular manifestations of Behçet's disease: a longitudinal study of up to 10 years. Int J Rheum Dis. 2014;17:444–52.
42. Ait Ben Haddou E, Imounan F, Regragui W, et al. Neurological manifestations of Behçet's disease: evaluation of 40 patients treated by cyclophosphamide. Rev Neurol Paris. 2012;168:344–9.
43. Noel N, Bernard R, Wechsler B, Resche-Rigon M, Depaz R, Le Thi Huong Boutin D, et al. Long-term outcome of neuro-Behçet's disease. Arthritis Rheumatol. 2014;66:1306–14.
44. Yoon D-L, Kim YJ, Koo BS, Kim Y-G, Lee C-K, Yoo B. Neuro-behçet's disease in South Korea: clinical characteristics and treatment response. Int J Rheum Dis. 2014;17:453–8.
45. Esatoglu SN, Hatemi G. Update on the treatment of Behçet's syndrome. Intern Emerg Med. 2019;14:661–75.
46. Yilmaz N, Emmungil H, Gucenmez S, Ozen G, Yildiz F, Balkarli A, et al. Incidence of cyclophosphamide-induced urotoxicity and protective effect of mesna in rheumatic diseases. J Rheumatol. 2015;42(9):1661–6.

47. Arida A, Fragiadaki K, Giavri E, Sfikakis PP. Anti-TNF agents for Behçet's disease: analysis of published data on 369 patients. Semin Arthritis Rheum. 2011;41(1):61–70.
48. Vallet H, Riviere S, Sanna A, Deroux A, Moulis G, Addimanda O, et al. Efficacy of anti-TNF alpha in severe and/or refractory Behçet's disease: multicenter study of 124 patients. J Autoimmun. 2015;62:67–74.
49. Adler S, Baumgartner I, Villiger PM. Behcet's disease: successful treatment with infliximab in 7 patients with severe vascular manifestations. A retrospective analysis. Arthritis Care Res (Hoboken). 2012;64:607–11.
50. Emmi G, Bettiol A, Silvestri E, Di Scala G, Becatti M, Fiorillo C, et al. Vascular Behçet's syndrome: an update. Intern Emerg Med. 2018;14:645–65.
51. Hamuryudan V, Seyahi E, Ugurlu S, Melikoglu M, Hatemi G, Ozguler Y, et al. Pulmonary artery involvement in Behçet's syndrome: effects of anti-TNF treatment. Semin Arthritis Rheum. 2015;45:369–73.
52. Desbois AC, Biard L, Addimanda O, Lambert M, Hachulla E, Launay D, Ackermann F, Perard L, Hot A, Maurier F, Mausservey C, Bernard F, Noel N, Alric L, Mirault T, Cohen F, Boussouar S, Resche-Rigon M, Cacoub P, Saadoun D. Efficacy of anti-TNF alpha in severe and refractory major vessel involvement of Behcet's disease: a multicenter observational study of 18 patients. Clin Immunol. 2018;197:54–9.
53. Sfikakis PP, Theodossiadis PG, Katsiari CG, Kaklamanis P, Markomichelakis NN. Effect of infliximab on sight-threatening panuveitis in Behçet's disease. Lancet. 2001;358:295–6.
54. Sfikakis PP, Kaklamanis PH, Elezoglou A, et al. Infliximab for recurrent, sight-threatening ocular inflammation in Adamantiades-Behcet disease. Ann Intern Med. 2004;140(5):404–6.
55. Cantini F, Niccoli L, Nannini C, et al. Efficacy of infliximab in refractory Behçet's disease-associated and idiopathic posterior segment uveitis: a prospective, follow-up study of 50 patients. Biologics. 2012;6(5):5–12.
56. Tugal-Tutkun I, Mudun A, Urgancioglu M, Kamali S, Kasapoglu E, Inanc M, et al. Efficacy of infliximab in the treatment of uveitis that is resistant to treatment with the combination of azathioprine, cyclosporine, and corticosteroids in Behçet's disease: an open-label trial. Arthritis Rheum. 2005;52:2478–84.
57. Fabiani C, Sota J, Rigante D, Vitale A, Emmi G, Lopalco G, Vannozzi L, Guerriero S, Bitossi A, Orlando I, Franceschini R, Frediani B, Galeazzi M, Iannone F, Tosi GM, Cantarini L. Efficacy of adalimumab and infliximab in recalcitrant retinal vasculitis inadequately responsive to other immunomodulatory therapies. Clin Rheumatol. 2018;37:2805–9.
58. Guzelant G, Uçar D, Esatoglu SN, et al. Infliximab for uveitis of Behçet's syndrome: a trend for earlier initiation. Clin Exp Rheumatol. 2017;35(Suppl. 108):S86–9.
59. Keino H, Okada AA, Watanabe T, Nakayama M, Nakamura T. Efficacy of infliximab for early remission induction in refractory uveoretinitis associated with Behçet's disease: a 2-year follow-up study. Ocul Immunol Inflamm. 2017;25:46–51.
60. Naganuma M, Sakuraba A, Hisamatsu T, et al. Efficacy of infliximab for induction and maintenance of remission in intestinal Behçet's disease. Inflamm Bowel Dis. 2008;14:1259–64.
61. Iwata S, Saito K, Yamaoka K, et al. Efficacy of combination therapy of anti-TNF-α antibody infliximab and methotrexate in refractory entero-Behçet's disease. Mod Rheumatol. 2011;21:184–91.
62. Lee JH, Cheon JH, Jeon SW, et al. Efficacy of infliximab in intestinal Behçet's disease: a Korean multicenter retrospective study. Inflamm Bowel Dis. 2013;19:1833–8.
63. Kinoshita H, Kunisaki R, Yamamoto H, et al. Efficacy of infliximab in patients with intestinal Behçet's disease refractory to conventional medication. Intern Med. 2013;52:1855–62.
64. Inoue N, Kobayashi K, Naganuma M, et al. Long-term safety and efficacy of adalimumab for intestinal Behçet's disease in the open-label study following a phase 3 clinical trial. Intest Res. 2017;15:395–401.
65. Licata G, Pinto A, Tuttolomondo A, Banco A, Ciccia F, Ferrante A, Triolo G. Anti-tumour necrosis factor-alpha monoclonal antibody therapy for recalcitrant cerebral vasculitis in a patient with Behcet's syndrome. Ann Rheum Dis. 2003;62:280–1.

66. Sarwar H, McGrath H Jr, Espinoza LR. Successful treatment of long-standing neuro-Behcet's disease with infliximab. J Rheumatol. 2005;32:181–3.
67. Ribi C, Sztajzel R, Delavelle J, Chizzolini C. Efficacy of TNF(alpha) blockade in cyclophosphamide resistant neuro-Behcet disease. J Neurol Neurosurg Psychiatry. 2005;76:1733–5.
68. Fujikawa K, Aratake K, Kawakami A, Aramaki T, Iwanaga N, Izumi Y, Arima K, Kamachi M, Tamai M, Huang M, Nakamura H, Nishiura Y, Origuchi T, Ida H, Eguchi K. Successful treatment of refractory neuro-Behcet's disease with infliximab: a case report to show its efficacy by magnetic resonance imaging, transcranial magnetic stimulation and cytokine profile. Ann Rheum Dis. 2007;66:136–7.
69. Alty JE, Monaghan TM, Bamford JM. A patient with neuro-Behcet's disease is successfully treated with etanercept: further evidence for the value of TNFalpha blockade. Clin Neurol Neurosurg. 2007;109:279–81.
70. Belzunegui J, Lopez L, Paniagua I, Intxausti JJ, Maiz O. Efficacy of infliximab and adalimumab in the treatment of a patient with severe neuro-Behcet's disease. Clin Exp Rheumatol. 2008;26:S133–4.
71. Madanat WY, Madanat AY. Worsening of osteomalacia in a 918 patient successfully treated for neuro-Behcet's disease with infliximab. Clin Exp Rheumatol. 2008;26:S128–9.
72. Abalos-Medina GM, Sanchez-Cano D, Ruiz-Villaverde G, Ruiz-Villaverde R, Quirosa Flores S, Raya Alvarez E. Successful use of infliximab in a patient with neuro-Behcet's disease. Int J Rheum Dis. 2009;12:264–6.
73. Borhani-Haghighi A, Safari A, Nazarinia MA, Habibagahi Z, Shenavandeh S. Infliximab for patients with neuro-Behcet's disease: case series and literature review. Clin Rheumatol. 2011;30(7):1007–12.
74. Fasano A, D'Agostino M, Caldarola G, Feliciani C, De Simone C. Infliximab monotherapy in neuro-Behcet's disease: four-year follow-up in a long-standing case resistant to conventional therapies. J Neuroimmunol. 2011;239:105–7.
75. Coulter I, Huda S, Baborie A, et al. Longitudinally extensive transverse myelitis as the sole presentation of neuro-Behçet's disease responding to infliximab. J Spinal Cord Med. 2012;35:122–4.
76. Kuroda R, Suzuki J, Muramatsu M, Tasaki A, Yano M, Imai N, et al. Efficacy of infliximab in neuro-Behçet's disease presenting with isolated longitudinally extensive transverse myelitis. J Neurol. 2013;260(12):3167–70.
77. Di Filippo M, Di Gregorio M, Nannini C, Gaetani L, Gallina C, Floridi P, et al. Infliximab monotherapy for neuro-Behçet's disease: a case report. J Neurol Sci. 2014;347(1–2):389–90.
78. Uygunoglu U, Pasha M, Saip S, Siva A. Recurrent longitudinal extensive transverse myelitis in a neuro-Behçet syndrome treated with infliximab. J Spinal Cord Med. 2015;38(1):111–4.
79. Barešić M, Reihl M, Habek M, Vukojević N, Anić B. Improvement of neurological and ocular symptoms of Behçet's disease after the introduction of infliximab. Rheumatol Int. 2018;38:1301–6.
80. Nigro A, Zuccoli G, Boiardi L, Salvarani C. Infliximab for the treatment of neuro-Behcet's disease: a case series and review of the literature. Arthritis Rheum. 2008;59:285–90.
81. Kikuchi H, Aramaki K, Hirohata S. Effect of infliximab in progressive Neuro-Behçet's syndrome. J Neurol Sci. 2008;272:99–105.
82. Matsui T, Ishida T, Tono T, Yoshida T, Sato S, Hirohata S. An attack of acute neuro-Behçet's disease during the course of chronic progressive neuro-Behçet's disease: report of two cases. Mod Rheumatol. 2010;20:621–6.
83. Al-araji A, Saip ASS, Constantinescu C, Akman-Demir G, Arayssi T, Espinoza L, Findling O, Garcia F, Hirohata S, Keogan M, Lo Monaco A, Pay S, Ramo C, Van Laar J, Zandi M. Treatment of neuro-Behcet's disease with infliximab. An International Multicentre Case-Series of 18 patients. Clin Exp Rheumatol. 2010;28:S119.
84. Zeydan B, Uygunoglu U, Saip S, et al. Infliximab is a plausible alternative for & neurologic complications of Behcet disease. Neurol Neuroimmunol Neuroinflamm. 2016;3:e258.
85. Giardina A, Ferrante A, Ciccia F, Vadalà M, Giardina E, Triolo G. One year study of efficacy and safety of infliximab in the treatment of patients with ocular and neurological Behçet's disease refractory to standard immunosuppressive drugs. Rheumatol Int. 2011;31:33–7.

86. Desbois AC, Addimanda O, Bertrand A, Deroux A, Pérard L, Depaz R, et al. Efficacy of anti-TNFα in severe and refractory neuro-Behcet disease. Medicine. 2016;95:e3550.
87. Hibi T, Hirohata S, Kikuchi H, et al. Infliximab therapy for intestinal, neurological, and vascular involvement in Behcet disease: efficacy, safety, and pharmacokinetics in a multicenter, prospective, open-label, single-arm phase 3 study. Medicine (Baltimore). 2016;95:e3863.
88. Sfikakis PP, Arida A, Panopoulos S, Fragiadaki K, Pentazos G, Laskari K, et al. Brief report: drug-free long-term remission in severe Behçet's disease following withdrawal of successful anti-tumorr necrosis factor treatment. Arthritis Rheumatol. 2017;69(12):2380–5.
89. Ryu YS, Park SH, Kim JM, Kim EJ, Lee J, Kwok SK, et al. A case of leukoencephalopathy associated with adalimumab-treated rheumatoid arthritis and a review of literature. Rheumatol Int. 2011;32(11):3481–5.
90. Melikoglu M, Fresko I, Mat C, et al. Short-term trial of etanercept in Behçet's disease: a double blind, placebo controlled study. J Rheumatol. 2005;32:98–105.
91. Atzeni F, Sarzi-Puttini P, Capsoni F, Mecchia M, Marazza MG, Carrabba M. Successful treatment of resistant Behcet's disease with etanercept. Clin Exp Rheumatol. 2005;23:729.
92. Scheinberg MA. Treatment of recurrent oral aphthous ulcers with etanercept. Clin Exp Rheumatol. 2005;23:733–4.
93. Sommer A, Altmeyer P, Kreuter A. A case of mucocutaneous Behçet's disease responding to etanercept. J Am Acad Dermatol. 2005;52:717–9.
94. Curigliano V, Giovinale M, Fonnesu C, et al. Efficacy of etanercept in the treatment of a patient with Behçet's disease. Clin Rheumatol. 2008;27:933–6.
95. Zhang MF, Zhao C, Wen X, et al. The short-term efficacy and safety treatment study of recurrent uveitis in Behçet's disease with etanercept. Zhonghua Yan Ke Za Zhi. 2010;46(2):145–50.
96. Ma D, Zhang CJ, Wang RP, Wang L, Yang H. Etanercept in the treatment of intestinal Behçet's disease. Cell Biochem Biophys. 2014;69:735–9.
97. Alty JE, Monaghan TM, Bamford JM. A patient with neuro-Behçet's disease is successfully treated with etanercept: further evidence for the value of TNFalpha blockade. Clin Neurol Neurosurg. 2007;109:279–81.
98. Sfikakis PP, Markomichelakis N, Alpsoy E, Assaad-Khalil S, Bodaghi B, Gul A, et al. Anti-TNF therapy in the management of Behçet's disease–review and basis for recommendations. Rheumatology (Oxford). 2007;46:736–41.
99. Cantarini L, Tinazzi I, Caramaschi P, Bellisai F, Brogna A, Galeazzi M. Safety and efficacy of etanercept in children with juvenile-onset Behçet's disease. Int J Immunopathol Pharmacol. 2009;22(2):551–5.
100. Saurenmann RK, Levin AV, Rose JB, et al. Tumour necrosis factor {alpha} inhibitors in the treatment of childhood uveitis. Rheumatology. 2006;45:982–9.
101. Jaffe GJ, Dick AD, Brézin AP, Nguyen QD, Thorne JE, Kestelyn P, et al. Adalimumab in patients with active noninfectious uveitis. N Engl J Med. 2016;375:932–43.
102. Nguyen QD, Merrill PT, Jaffe GJ, Dick AD, Kurup SK, Sheppard J, et al. Adalimumab for prevention of uveitic flare in patients with inactive non-infectious uveitis controlled by corticosteroids (VISUAL II): a multicentre, double-masked, randomized, placebo-controlled phase 3 trial. Lancet. 2016;388:1183–92.
103. Fabiani C, Vitale A, Emmi G, Vannozzi L, Lopalco G, Guerriero S, et al. Efficacy and safety of adalimumab in Behçet's disease-related uveitis: a multicenter retrospective observational study. Clin Rheumatol. 2017;36:183–9.
104. Fabiani C, Sota J, Vitale A, et al. Cumulative retention rate of adalimumab in patients with Behçet's disease-related uveitis: a four-year follow-up study. Br J Ophthalmol. 2018;102:637–41.
105. Fabiani C, Vitale A, Emmi G, et al. Long- term retention rates of adalimumab and infliximab in non-infectious intermediate, posterior and panuveitis. Clin Rheumatol. 2018;38:63–70.
106. Bitossi A, Bettiol A, Silvestri E, Di Scala G, Bacherini D, Lopalco G, et al. Adalimumab accounts for long-term control of non-infectious uveitis also in the absence of concomitant DMARD treatment: a multicenter retrospective study. Mediat Inflamm. 2019;2019:1623847.

107. Atienza-Mateo B, Martin-Varillas JL, Calvo-Rio V, et al. Comparative study of infliximab versus adalimumab in refractory uveitis due to Behçet's disease. National multicenter study of 177 cases. Arthritis Rheumatol. 2019;71(12):2081–9.
108. Fabiani C, Vitale A, Rigante D, et al. Comparative efficacy between adalimumab and infliximab in the treatment of non-infectious intermediate uveitis, posterior uveitis, and panuveitis: a retrospective observational study of 107 patients. Clin Rheumatol. 2019;38:407–15.
109. Emmi G, Vitale A, Silvestri E, et al. Adalimumab-based treatment versus disease-modifying antirheumatic drugs for venous thrombosis in Behcet's syndrome. Arthritis Rheumatol. 2018;70:1500–7.
110. Tanida S, Inoue N, Kobayashi K, et al. Adalimumab for the treatment of Japanese patients with intestinal Behçet's disease. Clin Gastroenterol Hepatol. 2015;13:940–8 e3.
111. Mizoshita T, Katano T, Tanida S, Hirano A, Miyaki T, Ozeki K, et al. Prospective comparison of preference and efficacy of adalimumab and infliximab for treating ulcerative colitis naive to anti-tumor necrosis factor therapy. Medicine. 2017;96:e7800.
112. Inoue N, Kobayashi K, Naganuma M, Hirai F, Ozawa M, Arikan D, et al. Long-term safety and efficacy of adalimumab for intestinal Behcet's disease in the open-label study following a phase 3 clinical trial. Intest Res. 2017;15:395–401.
113. Belzunegui J, Lopez L, Paniagua I, et al. Efficacy of infliximab and adalimumab in the treatment of a patient with severe neuro-Behçet's disease. Clin Exp Rheumatol. 2008;26:133–4.
114. Bawazeer A, Raffa L, Nizamuddin S. Clinical experience with adalimumab in the treatment of ocular Behçet's disease. Ocul Immunol Inflamm. 2010;18:226–32.
115. Olivieri I, D'Angelo S, Padula A, et al. Successful treatment of recalcitrant genital ulcers of Behçet's disease with adalimumab after failure of infliximab and etanercept. Clin Exp Rheumatol. 2009;27:112.
116. Lee S, Lee S, Kim K, et al. Adalimumab treatment for life threatening pulmonary artery aneurysm in Behcet disease: a case report. Clin Rheumatol. 2010;29:91–3.
117. Fox JD, Baquerizo-Nole KL, Keegan BR, Macquhae F, Escandon J, Espinosa A, Kirsner RS. Adalimumab treatment leads to reduction of tissue tumor necrosis factor-alpha correlated with venous leg ulcer improvement: a pilot study. Int Wound J. 2015;13(5):963–6.
118. Interlandi E, Leccese P, Olivieri I, Latanza L. Adalimumab for treatment of severe Behçet's uveitis: a retrospective long-term follow-up study. Clin Exp Rheumatol. 2014;32:S58–62.
119. Nanthapisal S, Klein NJ, Ambrose N, Eleftheriou D, Brogan PA. Paediatric Behçet's disease: a UK tertiary centre experience. Clin Rheumatol. 2016;35:2509–16.
120. Lim KI, Yang DH, Ryoo E. Behçet's disease with multiple splenic abscesses in a child. Intest Res. 2017;15:422–8.
121. Atienza-Mateo B, Calvo-Río V, Beltrán E, et al. Anti-interleukin 6 receptor tocilizumab in refractory uveitis associated with Behçet's disease: multicentre retrospective study. Rheumatology (Oxford). 2018;57:856–64.
122. Robinson AB, Gallentine WB, Rabinovich CE. Pediatric neuro-Behçet's disease responsive to adalimumab. Pediatr Neurol. 2010;43(4):291–3.
123. Marsili M, Marzetti V, Lucantoni M, Lapergola G, Gattorno M, Chiarelli F, Breda L. Autoimmune sensorineural hearing loss as presenting manifestation of paediatric Behçet disease responding to adalimumab: a case report. Ital J Pediatr. 2016;42:81.
124. Ozguler Y, Leccese P, Christensen R, Esatoglu SN, Bang D, Bodaghi B, Celik AF, Fortune F, Gaudric J, Gul A, Kotter I, Mahr A, Moots RJ, Richter J, Saadoun D, Salvarani C, Scuderi F, Sfikakis PP, Siva A, Stanford M, Tugal-Tutkun I, West R, Yurdakul S, Olivieri I, Yazici H, Hatemi G. Management of major organ involvement of Behcet's syndrome: a systematic review for update of the EULAR recommendations. Rheumatology (Oxford). 2018 Dec 1;57(12):2200–12.
125. Botsios C, Sfriso P, Furlan A, et al. Resistant Behçet disease responsive to anakinra. Ann Intern Med. 2008;149:284–6.
126. Cantarini L, Vitale A, Scalini P, Dinarello CA, Rigante D, Franceschini R, et al. Anakinra treatment in drug-resistant Behcet's disease: a case series. Clin Rheumatol. 2015;34:1293–301.

127. Cantarini L, Talarico R, Generali E, et al. Safety profile of biologic agents for Behçet's disease in a multicenter observational cohort study. Int J Rheum Dis. 2017;20(1):103–8.
128. Emmi G, Talarico R, Lopalco G, Cimaz R, Cantini F, Viapiana O, et al. Efficacy and safety profile of anti-interleukin-1 treatment in Behcet's disease: a multicenter retrospective study. Clin Rheumatol. 2016;35(5):1281–6.
129. Grayson PC, Yazici Y, Merideth M, Sen HN, Davis M, Novakovich E, et al. Treatment of mucocutaneous manifestations in Behcet's disease with anakinra: a pilot open-label study. Arthritis Res Ther. 2017;19(1):69.
130. Fabiani C, Vitale A, Emmi G, Lopalco G, Vannozzi L, Guer-riero S, Gentileschi S, Bacherini D, Franceschini R, Frediani B, Galeazzi M, Iannone F, Tosi GM, Cantarini L. Interleukin (IL)-1 inhibition with anakinra and canakinumab in Behcet's disease-related uveitis: a multicenter retrospective observational study. Ocul Immunol Inflamm. 2020;28(2):298–304.
131. Fabiani C, Vitale A, Rigante D, Emmi G, Lopalco G, Sota J, Vannozzi L, di Scala G, Guerriero S, Orlando I, Franceschini R. The presence of uveitis is associated with a sustained response to the interleukin (IL)-1 inhibitors anakinra and canakinumab in Behçet's disease. Ocul Immunol Inflamm. 2018;28:1–7. https://doi.org/10.1080/09273948.2018.1511810.
132. Caso F, Rigante D, Vitale A, et al. Efficacy of anakinra in refractory Behçet's disease sacroiliitis. Clin Exp Rheumatol. 2014;32:S171.
133. Vitale A, Rigante D, Caso F, et al. Interleukin-1 blockade in neuro-Behçet's disease: a case-based reflection. Int J Rheum Dis. 2017;20(8):1046–9.
134. Ugurlu S, Ucar D, Seyahi E, Hatemi G, Yurdakul S. Canakinumab in a patient with juvenile Behcet's syndrome with refractory eye disease. Ann Rheum Dis. 2012;71:1589–91.
135. Emmi G, Silvestri E, Ciucciarelli L, Squatrito D, Emmi L. Reply: anti-IL1 blocking agents in drug-resistant Behçet's syndrome: our little case series. Clin Exp Rheumatol. 2014;32:S172.
136. Vitale A, Rigante D, Caso F, et al. Inhibition of interleukin-1by canakinumab as a successful mono-drug strategy for the treatment of refractory Behçet's disease: a case series. Dermatology. 2014;228:211–4.
137. Pagnini I, Bondi T, Simonini G, Giani T, Marino A, Cimaz R. Successful treatment with canakinumab of a paediatric patient with resistant Behçet's disease. Rheumatology (Oxford). 2015;54:1327–8.
138. Emmi G, Talarico R, Lopalco G, et al. Efficacy and safety profile of anti- interleukin-1 treatment in Behçet's disease: a multicenter retrospective study. Clin Rheumatol. 2016;35(5):1281–6.
139. Orlando I, Vitale A, Rigante D, et al. Long-term efficacy and safety of the interleukin-1 inhibitors anakinra and canakinumab in refractory Behçet disease uveitis and concomitant bladder papillary carcinoma. Intern Med J. 2017;47:1086–8.
140. Tanaka T, Narazaki M, Ogata A, Kishimoto T. A new era for the treatment of inflammatory autoimmune diseases by interleukin-6 blockade strategy. Semin Immunol. 2014;26:88–96.
141. Hirano T, Ohguro N, Hohki S, Hagihara K, Shima Y, Narazaki M, et al. A case of Behçet's disease treated with a humanized anti-interleukin-6 receptor antibody, tocilizumab. Mod Rheumatol. 2012;22:298–302.
142. Calvo-Rio V, de la Hera D, Beltran-Catalan E, et al. Tocilizumab in uveitis refractory to other biologic drugs: a study of 3 cases and a literature review. Clin Exp Rheumatol. 2014;32(4 Suppl 84):S54–7.
143. Papo M, Bielefeld P, Vallet H, et al. Tocilizumab in severe and refractory non-infectious uveitis. Clin Exp Rheumatol. 2014;32:S75–9.
144. Deroux A, Chiquet C, Bouillet L. Tocilizumab in severe and refractory Behçet's disease: four cases and literature review. Semin Arthritis Rheum. 2016;45:733–7.
145. Alokaily F, Al Saati A, Jawad A. Successful treatment of Behçet's uveitis with Tocilizumab. Saudi J Ophthalmol. 2017;31(1):42–4.
146. Ozturk HE, Oray M, Tugal-Tutkun I. Tocilizumab for the treatment of Behçet uveitis that failed interferon alpha and anti-tumor necrosis factor alpha therapy. Ocul Immunol Inflamm. 2018;26:1005–14.

147. Atienza-Mateo B, Calvo-Río V, Beltrán E, Martínez-Costa L, Valls-Pascual E, Hernández-Garfella M, et al. Anti-interleukin 6 receptor tocilizumab in refractory uveitis associated with Behçet's disease: multicentre retrospective study. Rheumatology. 2018;57:856–64.
148. Diamantopoulos AP, Hatemi G. Lack of efficacy of tocilizumab in mucocutaneous Behçet's syndrome: report of two cases. Rheumatology (Oxford). 2013;52:1923–4.
149. Cantarini L, Lopalco G, Vitale A, et al. Paradoxical mucocutaneous flare in a case of Behçet's disease treated with tocilizumab. Clin Rheumatol. 2015;34:1141–3.
150. Emmi G, Silvestri E, Squatrito D, Emmi L, Cantarini L, Prisco D. Tocilizumab-induced exacerbation of mucosal ulcers in a patient with multi-refractory Behçet's disease. Semin Arthritis Rheum. 2016;46(1):e1–2.
151. Ding Y, Li C, Liu J, et al. Tocilizumab in the treatment of severe and/or refractory vasculo-Behçet's disease: a single-centre experience in China. Rheumatology (Oxford). 2018;57:2057–9.
152. Hirohata S, et al. Cerebrospinal fluid interleukin-6 in progressive Neuro-Behçet's syndrome. Clin Immunol Immunopathol. 1997;82:12–7.
153. Akman-Demir G, et al. Interleukin-6 in neuro-Behçet's disease: association with disease subsets and long-term outcome. Cytokine. 2008;44:373–6.
154. Urbaniak P, Hasler P, Kretzschmar S. Refractory neuro-Behçet treated by tocilizumab: a case report. Clin Exp Rheumatol. 2012;30(3 Suppl 72):S73–5.
155. Shapiro LS, Farrell J, Borhani Haghighi A. Tocilizumab treatment for neuro-Behçet's disease, the first report. Clin Neurol Neurosurg. 2012;114:297–8.
156. Addimanda O, Pipitone N, Pazzola G, Salvarani C. Tocilizumab for severe refractory neuro-Behçet: three cases IL-6 blockade in neuro-Behçet. Semin Arthritis Rheum. 2015;44:472–5.
157. Hatemi G, Seyahi E, Fresko I, Talarico R, Hamuryudan V. One year in review 2019: Behçet's syndrome. Clin Exp Rheumatol. 2019;37 Suppl 121(6):3–17.
158. Alpsoy E, Durusoy C, Yilmaz E, Ozgurel Y, Ermis O, Yazar S, et al. Interferon alfa-2a in the treatment of Behçet disease: a randomized placebo-controlled and double-blind study. Arch Dermatol. 2002;138:467–71.
159. Azizlerli G, Sarica R, Köse A, Ovül C, Kavala M, Kayabali M, et al. Interferon alfa-2a in the treatment of Behçet's disease. Dermatology. 1996;192:239–41.
160. Tsambaos D, Eichelberg D, Goos M. Behçet's syndrome: treatment with recombinant leukocyte alpha-interferon. Arch Dermatol Res. 1986;278(4):335–6.
161. Monastirli A, Chroni E, Georgiou S, Ellul J, Pasmatzi E, Papathanasopoulos P, Tsambaos D. Interferon-alpha treatment for acute myelitis and intestinal involvement in severe Behcet's disease. QJM. 2010;103(10):787–90.
162. Yalçındag N, Köse HC. Comparison of the treatment results for Behçet uveitis in patients treated with infliximab and interferon. Ocul Immunol Inflamm. 2019;3:1–10.
163. Nichols JC, Ince A, Akduman L, Mann ES. Interferon-alpha 2a treatment of neuro-Behçet disease. J Neuroophthalmol. 2001;21(2):109–11.
164. Chroni E, Tsambaos D. Interferon-alpha for the treatment of severe entero- and neuro-Behçet's disease. J Dermatol. 2015;42(2):229–30.
165. Calguneri M, Onat AM, Ozturk MA, Ozcakar L, Ureten K, Akdogan A, Ertenli I, Kiraz S. Transverse myelitis in a patient with Behcet's disease: favorable outcome with a combination of interferon-alpha. Clin Rheumatol. 2005;24(1):64–6.
166. Calguneri M, Ozturk MA, Ertenli I, Kiraz S, Apras S, Ozbalkan Z. Effects of interferon-alpha treatment on the clinical course of refractory Behcet's disease: an open study. Ann Rheum Dis. 2003;62(5):492–3.
167. Shi J, Zhao C, Zhou J, et al. Effectiveness and safety of interferon α2a as an add-on treatment for refractory Behçet's uveitis. Ther Adv Chronic Dis. 2019;10:1–9.
168. Yang DS, Taylor SR, Lightman SL. Interferon-alpha in the Management of Patients with Behçet's disease. Br J Hosp Med (Lond). 2005;69(2008):575–9.
169. Imrie FR, Dick AD. Biologics in the treatment of uveitis. Curr Opin Ophthalmol. 2007;18(6):481–6.

170. Hensel M, Breitbart A, Ho AD. Autologous hematopoietic stem-cell transplantation for Behcet's disease with pulmonary involvement. N Engl J Med. 2001;344:69.
171. Marmont AM, Gualandi F, Van Lint MT, et al. Autologous hematopoietic stem cell transplant for refractory Behçet's disease. Paper presented at Bone Marrow Transplant. Oral presentation number O157, 2003.
172. Marmont AM, Gualandi F, Piaggio G, et al. Allogeneic bone marrow transplantation (BMT) for refractory Behçet's diseasee with severe CNS involvement. Bone Marrow Transplant. 2006;37:1061–3.
173. Yamato K. Successful cord blood stem cell transplantation for myelodysplastic syndrome with Behcet disease. Int J Hematol. 2003;77:82–5.
174. Rossi G, Moretta A, Locatelli F. Autologous hematopoietic stem cell transplantation for severe/refractory intestinal Behcet disease. Blood. 2004;103:748–50.
175. Tomonari A, Tojo A, Takahashi T, et al. Resolution of Behçet's disease after HLA-mismatched unrelated cord blood transplantation for myelodysplastic syndrome. Ann Hematol. 2004;83:464–6.
176. Maurer B, Hensel M, Max R, et al. Autologous hematopoietic stem cell transplantation for Behçet's disease with pulmonary involvement: analysis after 5 years of follow up. Ann Rheum Dis. 2006;65:127–9.
177. Daikeler T, Hugle T, Farge D, et al. Allogeneic hematopoietic SCT for patients with autoimmune diseases. Bone Marrow Transplant. 2009;44:27–33.
178. De Cata A, Intiso D, Bernal M, et al. Prolonged remission of neuro-Behcet disease following autologous transplantation. Int J Immunopathol Pharmacol. 2007;20:91–6.
179. Nonami A, Takenaka K, Sumida C, et al. Successful treatment of myelodysplastic syndrome (MDS)-related intestinal Behcet's disease by up-front cord blood transplantation. Intern Med. 2007;46:1753–6.
180. Chauhan S, Olujohungbe A, Moots RJ. Treatment of Behcet's syndrome and plasmacytoma in a patient with peripheral blood stem cell transplantation. Leuk Lymphoma. 2008;49:2377–9.
181. Statkute L, Oyama Y, Barr WG, et al. Autologous non-myeloablative hematopoietic stem cell transplantation for refractory systemic vasculitis. Ann Rheum Dis. 2008;67:991–7.
182. Daikeler T, Kotter I, Tyndall CB, Attarbaschi A, Apperley J, Guardiola P, et al. Haematopoietic stem cell transplantation for vasculitis including Behcet's disease and polychondritis – a retrospective analysis of patients recorded to the European Bone Marrow Transplantation (EBMT) and European League Against Rheumatism (EULAR) databases and a review of the literature. Ann Rheum Dis. 2007;66:202–7.
183. Lim SH, Hulsey M, Esler WV. Resolution of Behcet's disease after non-myeloablative allogeneic stem cell transplant for acute myeloid leukemia. Rheumatology. 2009;48:88–9.
184. Gooley TA, Chien JW, Pergam SA, et al. Reduced mortality after allogeneic hematopoietic-cell transplantation. N Engl J Med. 2010;363:2091–101.
185. Soysal T, Salihoğlu A, Esatoğlu SN, Gültürk E, Eşkazan AE, Hatemi G, et al. Bonemarrow transplantation for Behçet's disease: a case report and systematic review of the literature. Rheumatology. 2014;53:1136–41.
186. Lee WS, Kang MH, Hwang JH, Oh YJ, Yoo WH. Allogeneic hematopoietic stem cell transplantation for refractory Behcet's disease with myelodysplastic syndrome: a case report. Int J Rheum Dis. 2017;20(11):1860–1.
187. Kürtüncü M, Aydın BN, Gündüz T, Ala S, Akman G, World Congress of Neurology Kyoto, WCN17-2801, 2017.

Prognosis of Neuro-Behçet's Syndrome

Nazire Pınar Acar-Özen and Aslı Tuncer

Behçet's Syndrome

Behçet's syndrome (BS) is a multisystemic and inflammatory vasculitis. The major features include skin lesions, recurrent painful oral aphthous ulcers, genital ulcerations, and either anterior or posterior uveitis. Furthermore, patients may also experience venous and arterial thrombosis, aneurysms, arthralgia, intestinal lesions, or epididymitis [38]. The frequency of these symptoms mainly depends on the ethnicity and gender of the patients [39].

Behçet's Syndrome Prognosis

Yazici and Kural-Seyahi et al. found that younger age at onset has less favorable outcome. In their 20-year outcome study with 387 patients, they observed that the major cause of mortality in Behçet's syndrome is major vessel involvement. They reported 50% of the patients with pulmonary artery aneurysm (n=24) deceased within one year [9]. While after a 10-year follow-up, 5-year survival was reported to be 62%. They stated that earlier onset of treatment can lead to better outcome [21, 37]. In their series, central nervous system involvement and heart diseases were the second leading causes of mortality. Renal involvement and neoplasms were detected to be other causes [21, 37]. Studies of the Pediatric Behçet Disease Study Group and the Italian investigators stated that the ocular and neurological involvements are the major reasons of mortality and morbidity [7, 19].

N. P. Acar-Özen · A. Tuncer (✉)
Hacettepe University, Faculty of Medicine, Department of Neurology, Ankara, Turkey
e-mail: akurne@hacettepe.edu.tr

E. Tüzün, M. Kürtüncü (eds.), *Neuro-Behçet's Disease*,
https://doi.org/10.1007/978-3-030-55273-2_13

Neuro-Behçet's Syndrome

Central nervous system (CNS) disease occurs in 5–25% of the patients with Behçet's syndrome. Kural-Seyahi et al. [21] reported the frequency of the neurologic disease as 11%. In another series, 5.3% of the 323 patients had neurological symptoms in 1-year period [29]. The prevalence of the neurological manifestations is 24.2% in the Japanese nationwide registration data of the 6627 Behçet patients [14]. In the Iraqi prospective study, the prevalence of neuro-Behçet's syndrome (NBS) is 14% [4]. A retrospective French series reported the NBS frequency as 14% [25].

Behçet's syndrome can affect both central and peripheral nervous systems. Central nervous system lesions can be parenchymal or nonparenchymal. Parenchymal syndrome can present as brain stem symptoms, cranial neuropathy, cerebellar or pyramidal dysfunction, spinal cord involvement, myelopathy, optic neuropathy, and cerebral symptoms such as encephalopathy, hemiparesis, hemisensory loss, seizures, dysphasia, and mental changes including cognitive and mood disturbances. Nonparenchymal syndrome involves vascular manifestations, especially cerebral venous thrombosis, intracranial hypertension syndrome, and acute meningeal syndrome [16].

Parenchymal disease could be either acute (acute meningoencephalitis) or progressive according to its treatment response. Acute NBS is usually self-limiting and responds well to corticosteroid therapy [30].

Headache is the most common neurological symptom among Behçet patients (up to 88.9%) [3]. Nearly 5% of the patients with headache is associated with neurological involvement [27]. Approximately half of the patients are referred to neurology clinic with motor symptoms [31]. Cerebellar or brain stem findings are not as common as pyramidal findings on the onset of NBS. Cognitive impairment is relatively rare but is a debilitating condition. Psychiatric symptoms are diverse. Common psychiatric symptoms are euphoria, loss of insight, disinhibition, indifference to their disease, psychomotor agitation or retardation, psychosis, and obsessive behaviors [3, 5, 30].

Adult neurological symptoms usually start after 3–5 years following the systemic onset [2, 3, 18, 31, 33]. Talarico et al. [32] showed in their 117-patient retrospective analysis that the onset of CNS involvement was in the first 10 years, with a higher incidence rate in the first 5 years.

Disability Scales for Neuro-Behçet's Syndrome

Several disability scales are used to define the prognosis of NBS. In their report, published in 1999, Akman-Demir et al. stated a disability status. They summarized being self-sufficient as "1," being physically dependent as "2," being mentally dependent as "3," being both mentally and physically dependent as "4," and death as "5" [2].

Expanded Disability Status Scale (EDSS) is another disability scale from 0 to 10. EDSS 6 represents the moderate disability in which a patient requires assistance in walking and daily life activities [23]. EDSS is generally used for multiple sclerosis disability evaluation. EDSS covers most of the functional systems in NBS apart from the visual disability. Hence, several authors chose EDSS for the prognostic evaluation [13, 31]. In NBS, uveitis caused most of the visual symptoms; therefore, Siva et al. eliminated visual problems from the original EDS scale [31].

Hirohata et al. evaluated the disability with the Steinbrocker functional classification, which is a scale for rheumatoid arthritis. Class 3 defines the patients who need assistance for their daily activities; class 4 defines the bedridden state. This scale does not include the neuropsychiatric and cognitive symptoms, which cause critical dependency [11]. Noel et al. [25], evaluated the disability with Rankin scale which defines stroke severity. It is a 0 to 6 scale, where 0 refers to normal examination, 6 refers to death, 4 represents the need for assistance at some level, and 5 is for bedridden patients. In another study, prognosis of neurological status was defined by bedridden state, dependency, or death [28].

In some of the studies, authors used specific scoring systems to evaluate Behçet's syndrome prognosis. Clinical severity score (CSS) is one of these scoring systems. This system assesses each symptom: 1 point for each mild symptom (oral aphthous lesions, genital ulcerations, skin lesions, headaches, etc.), 2 points for each moderate symptom (arthritis, deep vein thrombosis, anterior uveitis, gastrointestinal bleeding, etc.), and 3 points for each severe disease manifestation (NBS, posterior or panuveitis, major vein thrombosis, arterial involvement, bowel perforation) [20]. International consensus recommendations for NBS highlight the use of Neuro-Behçet's disability score (NBDS) [16]. This scoring system is proposed by Kürtüncü et al., and it evaluates motor and cognitive functions ranging from 0 to 8 [22].

Prognosis of Different Subgroups

Prognosis of Parenchymal Neuro-Behçet's Syndrome

Akman-Demir et al. retrospectively evaluated 200 patients with neurological involvement. They classified the disease into subgroups as primary progressive, secondary progressive, and as silent neurological involvement. One hundred sixty-two patients had parenchymal CNS involvement, whereas the rest of the patients had vascular involvement. Sixty-seven percent of patients had a course with a relapse. The rest of them had slowly progressive form. The median number of the attacks was 1.5 in 6 years. One-third of the patients had severe sequela. During the relapses, 19% were dependent on another person either physically or mentally. The median time for dependency or death is 115.7 months. The univariate analysis showed that brain stem involvement, having two or more attacks, dependency at admittance, relapse during corticosteroid tapering, progressive course, and abnormal CSF findings like elevated protein levels and pleocytosis were related with poor prognosis [2]. In another study, this group also reported a 7-year prognosis of 42 patients. They stated that 67% of 27 patients initially presenting with headache and normal neurological examination had still normal neurological examination. In the follow-up, 26% of these patients had no attacks but either minor neurological findings or abnormal neuropsychological or electrophysiological test results. The lumbar punctures were performed in the first attack. The CSF findings of all (n = 5) progressive NBS patients (33% of all NBS patients) revealed high protein levels and pleocytosis. This could be related with poor prognosis. Uveitis was significantly higher in the NBS group than the headache-only group [1] (Table 1).

Siva et al. reported the characteristics of 107 neuroimaging studies of 164 patients and 72.1% had parenchymal involvement. After a 10-year follow-up, the survival rate was 95.7 ± 2.1%; 45.1% of all NBS patients had an EDSS score of 6 or more at the end of the follow-up period. They summarized that the duration of the neurological disease, progressive disease course, dysarthria, cerebellar symptoms other than dysarthria, and motor symptoms at onset were poor prognostic factors. Factors with favorable outcome were disease course limited to a single episode and headache at onset. Headache at onset was negatively correlated with progressive disease. Similar with the data of Akman- Demir, they found headache to be positively correlated with the course limited to a single episode [31].

In their retrospective analysis of 275 NBS patients, Noel et al. included 115 patients with only parenchymal involvement. They classified the group as acute (n = 78) and progressive (n = 37) NBS. For acute patients, they formed two subgroups as "single episode" and "relapsing-remitting form." For the progressive NBS group, they made a classification as "primary progressive form" and "progressive form with further relapses." They evaluated the disability with Rankin score. In acute NBS patients, 47% had only one episode, and 21% had further relapsing NBS. In progressive form, 20% had a primary progressive course, and 12% had further attacks [25]. In their series, patients presenting with a progressive neurologic course were found to be older at the time of diagnosis for both BS and NBS. Between acute and progressive onset, no difference was found either for mucosal or systemic symptoms and for geographic origin. Confusion occurred more frequently at the onset of progressive course, whereas the frequency of meningitis was similar for both groups [25]. In a follow-up of nearly 6 years of these 115 patients, 33% experienced at least one neurologic relapse. They state that the presence of HLA-B51 antigen and coma as a presenting symptom can be associated with risk of NBD relapse. However, in multivariant analysis, only HLA-B51 antigen was found to be related with NBD relapse. Noel et al. defined poor outcome as inability to perform activities of daily living and/or death. In univariate analysis, longer time to NBD diagnosis, the presence of hemiparesis or paraparesis, sensory symptoms, sphincter dysfunction, and poor baseline disability were related with poor outcome. In multivariate analysis, the presence of baseline hemiparesis or paraparesis and brain stem lesions were independently associated with poor outcome [25].

Hirohata et al. evaluated 37 chronic progressive NBS (CPNBS) cases with their clinical data, magnetic resonance imaging, and cerebrospinal fluid findings and with

Table 1 Review of the literature of neuro-behçet's syndrome prognosis

Author	Year	Number of NBS patients	Follow-up years (median)	Better prognosis	Poor prognosis
Akman-Demir et al.	1999	200	6		Abnormal CSF findings like elevated protein levels and pleocytosis, brain stem involvement, having 2 or more attacks, dependency at admittance, relapse during corticosteroid tapering, and progressive course
Akman-Demir et al.	1996	42	7	Initially presenting with headache and normal neurological examination	High protein levels and pleocytosis
Siva et al.	2001	164	10	Disease course limited to a single episode and headache at onset	Duration of neurological disease, progressive disease course, dysarthria, cerebellar symptoms other than dysarthria, and motor symptoms at onset
Noel et al.	2014	275	6		Longer time to NBD diagnosis, the presence of hemiparesis or paraparesis, sensory symptoms, sphincter dysfunction, and poor baseline disability In multivariate analysis: the presence of baseline hemiparesis or paraparesis Brain stem lesions HLA- B51 positivity related with relapse rate
Hirohata et al.	2015	37	16	Methotrexate use	Brain stem atrophy, higher CSF IL-6 levels
Sbai et al.	2003	109	8	Continuous treatment	Brain stem and internal capsule lesions and rhombencephalitis
Joseph et al.	2007	22	10		Repeated attacks, incomplete recovery, progressive disease course, extensive or spinal involvement, early neurological involvement, and CSF pleocytosis
Farahangiz et al.	2012	58	3.6		Initial brain stem atrophy
Houman et al.	2013	121	3		Male gender CNS parenchymal lesions
Gerber et al.	1996	12	3.5		New small lesions Increase in the atrophy

(continued)

Table 1 (continued)

Author	Year	Number of NBS patients	Follow-up years (median)	Better prognosis	Poor prognosis
Kalra et al.	2014	Diagnosis and management of neuro-Behçet's disease: International consensus recommendations			Brain stem and spinal cord presentation, frequent relapses, early disease progression, high CSF pleocytosis, disability and dependent status at initial presentation, a primary or secondary progressive course, relapse during steroid dose tapering, fever, meningeal signs, and bladder involvement

interleukin-6 levels. The patients were under corticosteroid and/or methotrexate treatment. After 100 and 200 months of the therapy initiation, the overall survival rate was 87.6% and 54.8%, respectively. They analyzed the cumulative ratios of either bedridden state or death. After 100 months of treatment, the ratio was 23.9% and increased to 65.4% after 200 months. The univariate and multivariate analyses revealed that only methotrexate use was associated with better prognosis. The CSF IL-6 levels were higher in nine Steinbrocker class 3–4 patients, whereas among the 19 Steinbrocker class 1 and 2 patients, IL-6 levels were lower. The authors emphasized the need for further prospective studies to clarify the relationship of the IL-6 with the NBS prognosis; 89.2% of the patients had brain stem atrophy in their magnetic resonance imaging (MRI) scans, which could be an indicator for the poor prognosis of the CPNBS patients [11]. In their retrospective series of 144 patients (acute NBS–CPNBS and non-NBS patients), they also highlighted that smoking was significantly higher among CPNBS group (91%) [10].

Sbai et al. included 109 parenchymal NBS cases in their retrospective study. They excluded the dural sinus thrombosis cases. Their parameters were relapses, attacks of other systems, CSF, imaging findings, and therapeutic choices. After the first neurological attack, the median follow-up period was 97 months; 45.8% of patients recovered well after their first attack, whereas 11% remained stable and 5.5% had a progressive course. Dependency ratio was 52% at the beginning of the study; after the follow-up period, 19% of the patients were physically and/or mentally dependent. Neurological disability was related to brain stem and internal capsule lesions and rhombencephalitis. Continuous treatment was associated with better outcomes compared with interrupted treatment. Continuous corticosteroid use was associated with less disability [28].

Joseph et al. from a district hospital in Bristol reviewed 22 Caucasian English of Welsh NBD patients. The mean follow-up period was 10 years (0.25–29.8 years). They regarded repeated attacks, incomplete recovery, progressive disease course, extensive or spinal involvement, and early neurological involvement as poor prognostic factors. CSF pleocytosis was also related either with a more severe disease or an attack [15].

Farahangiz et al. [6] from Iran reviewed the MRI characteristics of 49 NBS patients. They divided the patients as monophasic (31%), polyphasic (27%), and progressive (20%) patients and headache attributed to BS patients (22%). The mean follow-up time was 3.6 (0.8–6.9) years. Forty percent of the patients with progressive course had brain stem atrophy on their first MRI evaluation. Thus, they related the initial brain stem atrophy with the progressive course.

A Tunisian center evaluated the data of 121 NBS patients among 430 patients with BS. Seventy-four patients had parenchymal NBS [12]. They associated male gender and CNS parenchymal lesions with a poorer prognosis.

International consensus recommendations in 2014 state brain stem and spinal cord presentation, frequent relapses, early disease progression, high CSF pleocytosis, disability and dependent status at initial presentation, a primary or secondary progressive course, relapse during steroid dose tapering, fever, meningeal signs, and bladder involvement are associated with poor prognosis [16]. Gender, presence of other systemic manifestations of BS, and age at onset do not have any influence over prognosis [16, 31].

Magnetic resonance imaging follow-up of 12 patients for 1.5–6 years (mean, 3.5 years) showed that half of the patients had new lesions and nine patients had cerebral atrophy [8]. They concluded that new small lesions and the increase in the atrophy could be prognostic signs for bad outcome.

Non-parenchymal Neuro-Behçet's Syndrome Prognosis

Prognosis of Cerebral Venous Thrombosis in Neuro-Behçet's Syndrome

Among all cerebral venous thrombosis cases (n = 182), Wasay et al. defined the poor prognostic factors as coma at presentation, being older than 60 years, intracerebral hemorrhage, and hypotension on admission and thrombosis of three major sinuses. They defined the good prognostic factors as age less than 45 years, thrombolytic treatment, and isolated transverse sinus thrombosis. Multivariate analysis narrowed the predictors. Poor prognostic factors are coma at presentation and intracerebral hemorrhage [34].

Yesilot et al. compared cerebral venous thrombosis cases under Behçet's syndrome with other etiologies. BS patients were mostly male, and the median age was younger. The onset of BS CVT group happened to be subacute or chronic, and they had better prognosis [40]. BS CVT group had fewer cortical infarcts, and this could be associated with the better prognosis of the BS CVT group. Wechsler et al.'s study investigated the MRI lesions among Behçet patients with neurological involvement [35] (n = 31, CVT patients: n = 10), and the same group also studied the long-term follow-up of 25 CVT patients [36]. Most of the cases had chronic onset. The initial symptom was intracranial hypertension. Neurological symptoms improved in

4 weeks. Uluduz et al. reviewed the pediatric CVT cases; 88.5% of their patients had cerebral venous sinus thrombosis (CVST). None of their CVST cases had cortical infarcts and had generally good outcome [33].

Prognosis of Pediatric Neuro-Behçet's Syndrome

Child-onset neuro-Behçet's syndrome is rarer than adult-onset NBS. According to Uluduz et al., 4% of the total NBS cases are child onset [33]. The onset of the first neurological symptom is approximately 2 years after the other systemic symptoms in the Pediatric Behçet Disease study group cohort [19], and it is 1.25 years in the Turkish series of Uluduz et al. [33]. Uluduz et al. evaluated the prognosis according to the initial response to treatment, and they stated the residual neurological deficit as bad outcome [33]. Their mean follow-up period was 6 ± 5 years. Two of 26 patients had residual neurologic symptoms after treatment, and one of them was independent. They explain this better prognosis with CVST being common in child-onset NBS [33].

Mora et al. reviewed 130 cases with neuro-ophthalmological features between the years 1971 and 2011 [24]. Eighty percent of the patients recovered after treatment. Seventeen percent of the patients had visual or neurological sequelae, and one patient died after the rupture of a cerebral aneurysm [17, 24].

Gender Effect on Prognosis

Male predominance is a known fact in Behçet's syndrome. Most of the NBS series also emphasized the male dominance. Several studies evaluated the prognostic effect of gender. Studies stated male gender does not affect prognosis [11, 25, 31]. Akman-Demir et al.'s 7-year follow-up series showed that male-female ratio in the headache group was 0.87, whereas it increased to 2.85 in the group with neurologic involvement [1].

Epidemiologic Effect on Prognosis

Epidemiologic data about Behçet's syndrome reveals that gastrointestinal symptoms are more common in the Far East (Japan) and skin findings like pathergy positivity are more common in Turkey and Mediterranean countries [41]. There is no isolated epidemiologic data about the severity of NBS syndrome in the literature.

In an Israeli study ($n = 100$; 66 Jews, 34 Arabs), no difference was found between the two groups with respect to the rate of major oral ulcers, genital ulcers, ocular disease, skin lesions, positive pathergy reaction, or vascular or neurological involvement.

Genetic Effects on Prognosis

Genetic susceptibility to BS has been a study topic for many years. As HLA-B51 positivity is more common among patients with BS (approximately 60%), some authors proposed the disease as an MHC-I-opathy [38].

Hirohota et al. [11] stated that positivity of HLA-B51 was 86% in chronic progressive NBS. Noel et al. [25] showed HLA-B51 positivity increased the odds of relapse by 3.6-fold. Relapse rate was 50% among HLA-B51-positive patients, whereas it was 21% among HLA-B51-negative patients [25].

Mortality–Morbidity in Neuro-Behçet's Syndrome

Mortality rate of patients with Behçet's syndrome is higher among younger men (14–24 years of age and 25–34 years of age), whereas older men (35–50 years of age) and women had a normal life span [26, 38]. Mortality is also increased in earlier phases of the disease (the first 7 years), especially among the ones with major organ involvement. Major causes of mortality were large vessel disease and parenchymal central nervous system (CNS) disease [38].

A Turkish group collected 20-year follow-up data of 387 patients. The mortality rate was 9.8%. Five male patients died due to severe neurological involvement with recurrent attacks. At the end of 20 years, 25.8% ($n = 57$) of male patients and 8.2% ($n = 10$) of female patients (overall 17.3% of patients) were dependent. Visual loss (76.1%) and CNS disease (7.5%) caused the dependency in most of the patients [21]. In a cohort of 817 patients of a French group, mortality rate was 5% ($n = 41$). Median follow-up period was 8 years; 26.9% of those patients had CNS disease. Among the 41 patients, five died (12.2% of the patients who died) because of NBS, and the cause of death of one patient was cerebral aneurysm. In their series, multivariate analysis found male sex, arterial involvement, genital ulceration, and high frequency of BS flares were significantly and independently associated with mortality [26].

There are a few studies addressing the mortality rates of NBS. In Akman-Demir's 15-patient series, the mortality rate was 20% in 7 years [1]. Another study of the same group reviewed 200 patients (155 male; 45 female). Akman-Demir et al. reviewed 200 patients, and 110 had a follow-up period of 3 and more years [2]. Among those 110 patients, 28 patients became either physically or mentally dependent, and 22 were deceased (mortality rate was 20%).

In Siva et al.'s 15 years of retrospective evaluation, mortality ratio was 5.5%. They performed the survival analysis from the onset of BS. Ten and 15-year survivals were 95.7 ± 2.1% and 88.5 ± 4.5%, respectively. All of the three patients who died because of neurological problems had cerebellopyramidal syndrome; 45.1% of the patients had moderate to severe disability 10 years after the disease onset. But the authors stated that due to the short follow-up period, standard error is high [31].

Mortality ratio of a French group was 10.4% among 115 NBS patients. Five of those patients died due to central nervous system-related causes [25].

References

1. Akman-Demir G, Baykan-Kurt B, Serdaroglu P, Gurvit H, Yurdakul S, Yazici H, Bahar S, Aktin E. Seven-year follow-up of neurologic involvement in Behcet syndrome. Arch Neurol. 1996;53:691–4.
2. Akman-Demir G, Serdaroglu P, Tasci B. Clinical patterns of neurological involvement in Behcet's disease: evaluation of 200 patients. The Neuro-Behcet Study Group. Brain. 1999;122(Pt 11):2171–82.
3. Al-Araji A, Kidd DP. Neuro-Behcet's disease: epidemiology, clinical characteristics, and management. Lancet Neurol. 2009;8:192–204.
4. Al-Araji A, Sharquie K, Al-Rawi Z. Prevalence and patterns of neurological involvement in Behcet's disease: a prospective study from Iraq. J Neurol Neurosurg Psychiatry. 2003;74:608–13.
5. Siva A, Hirohata S. Behçet's Syndrome and the Nervous System. In Yazici Y, Yazici H (ed.), Behçet's Syndrome, 1st ed. (pp 95–114), Springer: New York, NY, USA.
6. Farahangiz S, Sarhadi S, Safari A, Borhani-Haghighi A. Magnetic resonance imaging findings and outcome of neuro-Behcet's disease: the predictive factors. Int J Rheum Dis. 2012;15:e142–9.
7. Gallizzi R, Pidone C, Cantarini L, Finetti M, Cattalini M, Filocamo G, Insalaco A, Rigante D, Consolini R, Maggio MC, Civino A, Martino S, Olivieri AN, Fabio G, Pastore S, Mauro A, Sutera D, Trimarchi G, Ruperto N, Gattorno M, Cimaz R. A national cohort study on pediatric Behcet's disease: cross-sectional data from an Italian registry. Pediatr Rheumatol Online J. 2017;15:84.
8. Gerber S, Biondi A, Dormont D, Wechsler B, Marsault C. Long-term MR follow-up of cerebral lesions in neuro-Behcet's disease. Neuroradiology. 1996;38:761–8.
9. Hamuryudan V, Yurdakul S, Moral F, Numan F, Tüzün H, Tüzüner N, Mat C, Tüzün Y, Ozyazgan Y, Yazïci H. Pulmonary arterial aneurysms in Behçet's syndrome: a report of 24 cases. Br J Rheumatol. 1994;33(1):48–51. https://doi.org/10.1093/rheumatology/33.1.48. PMID: 8162457.
10. Hirohata S, Kikuchi H, Sawada T, Nagafuchi H, Kuwana M, Takeno M, Ishigatsubo Y. Clinical characteristics of neuro-Behcet's disease in Japan: a multicenter retrospective analysis. Mod Rheumatol. 2012;22:405–13.
11. Hirohata S, Kikuchi H, Sawada T, Nagafuchi H, Kuwana M, Takeno M, Ishigatsubo Y. Retrospective analysis of long-term outcome of chronic progressive neurological manifestations in Behcet's disease. J Neurol Sci. 2015;349:143–8.
12. Houman MH, Bellakhal S, Ben Salem T, Hamzaoui A, Braham A, Lamloum M, Monia SK, Ben Ghorbel I. Characteristics of neurological manifestations of Behcet's disease: a retrospective monocentric study in Tunisia. Clin Neurol Neurosurg. 2013;115:2015–8.
13. Ideguchi H, Suda A, Takeno M, Kirino Y, Ihata A, Ueda A, Ohno S, Baba Y, Kuroiwa Y, Ishigatsubo Y. Neurological manifestations of Behcet's disease in Japan: a study of 54 patients. J Neurol. 2010;257:1012–20.
14. Ishido T, Horita N, Takeuchi M, Kawagoe T, Shibuya E, Yamane T, Hayashi T, Meguro A, Ishido M, Minegishi K, Yoshimi R, Kirino Y, Kato S, Arimoto J, Ishigatsubo Y, Takeno M, Kurosawa M, Kaneko T, Mizuki N. Clinical manifestations of Behcet's disease depending on sex and age: results from Japanese nationwide registration. Rheumatology (Oxford). 2017;56:1918–27.
15. Joseph FG, Scolding NJ. Neuro-Behcet's disease in Caucasians: a study of 22 patients. Eur J Neurol. 2007;14:174–80.

16. Kalra S, Silman A, Akman-Demir G, Bohlega S, Borhani-Haghighi A, Constantinescu CS, Houman H, Mahr A, Salvarani C, Sfikakis PP, Siva A, Al-Araji A. Diagnosis and management of Neuro-Behcet's disease: international consensus recommendations. J Neurol. 2014;261:1662–76.
17. Kerr JS, Roach ES, Sinal SH, McWhorter JM. Intracranial arterial aneurysms complicating Behcet's disease. J Child Neurol. 1989;4:147–9.
18. Kidd D, Steuer A, Denman AM, Rudge P. Neurological complications in Behcet's syndrome. Brain. 1999;122(Pt 11):2183–94.
19. Kone-Paut I, Shahram F, Darce-Bello M, Cantarini L, Cimaz R, Gattorno M, Anton J, Hofer M, Chkirate B, Bouayed K, Tugal-Tutkun I, Kuemmerle-Deschner J, Agostini H, Federici S, Arnoux A, Piedvache C, Ozen S, PEDBD Group. Consensus classification criteria for paediatric Behcet's disease from a prospective observational cohort: PEDBD. Ann Rheum Dis. 2016;75:958–64.
20. Krause I, Mader R, Sulkes J, Paul M, Uziel Y, Adawi M, Weinberger A. Behcet's disease in Israel: the influence of ethnic origin on disease expression and severity. J Rheumatol. 2001;28:1033–6.
21. Kural-Seyahi E, Fresko I, Seyahi N, Ozyazgan Y, Mat C, Hamuryudan V, Yurdakul S, Yazici H. The long-term mortality and morbidity of Behcet syndrome: a 2-decade outcome survey of 387 patients followed at a dedicated center. Medicine (Baltimore). 2003;82:60–76.
22. Kurtuncu M, Tuzun E, Mutlu M, Pehlivan M, Serdaroglu R, Akman Demir G. Clinical patterns and course of neuro-Behçet's disease: analysis of 354 patients comparing cases presented before and after 1990. Clin Exp Rheumatol. 2008;26:S43.
23. Kurtzke JF. Rating neurologic impairment in multiple sclerosis: an expanded disability status scale (EDSS). Neurology. 1983;33:1444–52.
24. Mora P, Menozzi C, Orsoni JG, Rubino P, Ruffini L, Carta A. Neuro-Behcet's disease in childhood: a focus on the neuro-ophthalmological features. Orphanet J Rare Dis. 2013;8:18.
25. Noel N, Bernard R, Wechsler B, Resche-Rigon M, Depaz R, Le Thi Huong Boutin D, Piette JC, Drier A, Dormont D, Cacoub P, Saadoun D. Long-term outcome of neuro-Behcet's disease. Arthritis Rheumatol. 2014;66:1306–14.
26. Saadoun D, Wechsler B, Desseaux K, Le Thi Huong D, Amoura Z, Resche-Rigon M, Cacoub P. Mortality in Behcet's disease. Arthritis Rheum. 2010;62:2806–12.
27. Saip S, Siva A, Altintas A, Kiyat A, Seyahi E, Hamuryudan V, Yazici H. Headache in Behcet's syndrome. Headache. 2005;45:911–9.
28. Sbai A, Wechsler B, Duhaut P, Du-Boutin LT, Amoura Z, Cacoub P, Godeau P, Piette JC. Neuro-Behcet's disease (isolated cerebral thrombophlebitis excluded). Clinical pattern, prognostic factors, treatment and long term follow-up. Adv Exp Med Biol. 2003;528:371–6.
29. Serdaroglu P, Yazici H, Ozdemir C, Yurdakul S, Bahar S, Aktin E. Neurologic involvement in Behcet's syndrome. A prospective study. Arch Neurol. 1989;46:265–9.
30. Siva A, Altintas A, Saip S. Behcet's syndrome and the nervous system. Curr Opin Neurol. 2004;17:347–57.
31. Siva A, Kantarci OH, Saip S, Altintas A, Hamuryudan V, Islak C, Kocer N, Yazici H. Behcet's disease: diagnostic and prognostic aspects of neurological involvement. J Neurol. 2001;248:95–103.
32. Talarico R, d'Ascanio A, Figus M, Stagnaro C, Ferrari C, Elefante E, Baldini C, Tani C, Mosca M, Bombardieri S. Behcet's disease: features of neurological involvement in a dedicated centre in Italy. Clin Exp Rheumatol. 2012;30:S69–72.
33. Uluduz D, Kurtuncu M, Yapici Z, Seyahi E, Kasapcopur O, Ozdogan H, Saip S, Akman-Demir G, Siva A. Clinical characteristics of pediatric-onset neuro-Behcet disease. Neurology. 2011;77:1900–5.
34. Wasay M, Bakshi R, Bobustuc G, Kojan S, Sheikh Z, Dai A, Cheema Z. Cerebral venous thrombosis: analysis of a multicenter cohort from the United States. J Stroke Cerebrovasc Dis. 2008;17:49–54.
35. Wechsler B, Dell'lsola B, Vidailhet M, Dormont D, Piette JC, Bletry O, Godeau P. MRI in 31 patients with Behcet's disease and neurological involvement: prospective study with clinical correlation. J Neurol Neurosurg Psychiatry. 1993;56:793–8.

36. Wechsler B, Vidailhet M, Piette JC, Bousser MG, Dell Isola B, Bletry O, Godeau P. Cerebral venous thrombosis in Behcet's disease: clinical study and long-term follow-up of 25 cases. Neurology. 1992;42:614–8.
37. Yazici H, Esen F. Mortality in Behcet's syndrome. Clin Exp Rheumatol. 2008;26:S138–40.
38. Yazici H, Seyahi E, Hatemi G, Yazici Y. Behcet syndrome: a contemporary view. Nat Rev Rheumatol. 2018;14:107–19.
39. Yazici H, Ugurlu S, Seyahi E. Behcet syndrome: is it one condition? Clin Rev Allergy Immunol. 2012;43:275–80.
40. Yesilot N, Bahar S, Yilmazer S, Mutlu M, Kurtuncu M, Tuncay R, Coban O, Akman-Demir G. Cerebral venous thrombosis in Behcet's disease compared to those associated with other etiologies. J Neurol. 2009;256:1134–42.
41. Yurdakul S, Yazici Y. Epidemiology of Behçet's syndrome and regional differences in disease expression. In: Hasan Y, Yusuf Y, editors. Behçet's syndrome. New York: Springer Verlag; 2010.

Index

E. Tüzün, M. Kürtüncü (eds.), *Neuro-Behçet's Disease*,
https://doi.org/10.1007/978-3-030-55273-2

Q

The manufacturer's authorised representative in the EU is Springer Nature Customer Service Centre GmbH, Europaplatz 3, 69115 Heidelberg, Germany. If you have any concerns regarding our products, please contact ProductSafety@springernature.com

Printed and bound by CPI Group (UK) Ltd, Croydon, CR0 4YY
30/09/2026
02225515-0015